# HOG WILD

## THE BATTLE FOR WORKERS' RIGHTS AT THE WORLD'S LARGEST SLAUGHTERHOUSE

Lynn Waltz

UNIVERSITY OF IOWA PRESS | IOWA CITY

University of Iowa Press, Iowa City 52242

Copyright © 2018 by the University of Iowa Press

www.uipress.uiowa.edu

Printed in the United States of America

Design by April Leidig

The University of Iowa Press is a member of Green Press
Initiative and is committed to preserving natural resources.

Printed on acid-free paper

ISBN 978-1-60938-585-9 (pbk)

ISBN 978-1-60938-586-6 (ebk)

Cataloging-in-Publication data is on file with the Library
of Congress.

*As always, for Sophie and Zoe*

# CONTENTS

Acknowledgments ix

Preface xi

Introduction 1

1 Joe Luter and Smithfield Foods 9

2 Cheap Labor Built on a Legacy of Slavery 19

3 Lots of Pigs, Lots of Poop, Lots of Politics, Lots of Pollution 25

4 The Plant Opens; The Work Is Beastly 35

5 The First Union Vote; The NLRB Investigates 51

6 Environment and Immigration; North Carolina Is Forever Altered 69

7 The Company Woman; Climbing the Ladder Has Its Costs 77

8 The Second Union Vote; More Firings 91

9 The Trial; A Surprise Witness 111

10 The Judge Rules 131

11 On the Road with Union Organizers 141

12 The Corporate Campaign; Basic Human Rights 149

13 Pressure Mounts on Harris Teeter and Paula Deen 167

14 Latino Workers Walk Off the Job; ICE Raids Cause Flight 183

15 Stockholders Salvo; Secret Talks; Stalemate 203

16 RICO Shocks; Both Sides Flinch 217

Epilogue 231

Notes 243

Bibliographic Essay 273

Index 277

# ACKNOWLEDGMENTS

SPECIAL THANKS TO Sherri Buffkin and Gene Bruskin for their courage in speaking out and ensuring this story could be told, and to Jasper Brown, Keith Ludlum, and others who generously gave of their time. Also, without the hard work of journalists, particularly those at the *Fayetteville Observer* and the *News & Observer,* who documented the actions of Smithfield Foods and Smithfield Packing, this book would not have been possible. Thanks also to the staff at the National Labor Relations Board for their labors in providing thousands of pages of documentation in searchable form.

To my primary editor at the University of Iowa Press, Catherine Cocks, I owe a great debt for her hours of patient counsel and editing to bring the manuscript along. Thanks also to editor Ranjit Arab for his help in making final editorial decisions and to Meredith Stabel and the staff at the press for all their hard work in bringing the manuscript to fruition. Special thanks to copyeditor Christine Gever, whose attention to detail is truly remarkable.

Thanks to videographer and editor Sherry DiBari, whose insights and eagle-eye view were invaluable, and to photographer and professor Dr. Michael DiBari Jr., without whose humor, generosity, and goodwill I could not have prevailed. Thanks also to journalism professor and writer Wayne J. Dawkins for his early read and steadily encouraging words.

Appreciation goes out to author and professor Michael Pearson at Old Dominion University and to author Blake Bailey for their guidance on the thesis that preceded this book. And to Buzz Bissinger, who altered my life's course into journalism many years ago.

To my daughters Sophie and Zoe, it is difficult to know how to extend thanks for the many hours stolen from evenings, weekends, holidays, and summers, but sincere thanks for the readings, edits, and advice, and most of all, for excelling independently all the while.

# PREFACE

MY INTRODUCTION TO Smithfield Foods and its questionable practices came in 2004, while on a freelance assignment for *Virginia Business*. The resulting article, published in 2005, covered the company's legal problems with the National Labor Relations Board (NLRB) over firing union supporters at its plant in Tar Heel, North Carolina, the largest slaughterhouse in the world. It also looked at Smithfield's continued appeals of NLRB rulings and its resistance to state and federal environmental laws, which resulted in a $12.6 million federal fine, the EPA's largest at that time. The article outlined environmental problems the company's factory farms and contract farms were causing in North Carolina and contextualized Smithfield's overall dominance and rapid growth within the industry.

For that piece, I interviewed environmental activist Robert F. Kennedy Jr., who was fighting Smithfield's expansion in Poland while also trying to stop its environmental violations at home and elsewhere abroad. Kennedy believes that Smithfield outsources the true cost of its business onto its workers and the citizens in terms of environmental cleanup costs and loss of quality of life. He says that Smithfield turned the state of North Carolina into a "company town."

In August of 2004, I had gone to company headquarters in Smithfield, Virginia, to interview Joseph Luter IV, then an executive vice president of Smithfield Foods, who told me that governmental and environmental restraints were limiting growth opportunities. I also spoke with Larry Pope, then president and chief operating officer of Smithfield, who said that the company was committed to the highest principles of food, environmental, and worker safety. Pope said several times that they were "not the bad guys"—in the United States or in Poland.

While there, I toured Smithfield's Gwaltney plant, the only time I've been in a slaughterhouse. The memories are visceral. The roar of the machinery. The deep cold. The hogs' heads floating by. The smocks, rubber shoes, gloves, and hairnets. The rows of workers with large, wickedly sharp knives whacking off slices of meat from huge slabs. The blood. And the smell. The smell that clung to my suit.

In 2005, I wrote a second article, this time for *Portfolio Weekly*. I went to the Smithfield Packing slaughterhouse in Tar Heel to see firsthand what was happening with the labor controversy there. I did ride-alongs with several union organizers and attended their morning meetings and public events. I interviewed local campaign organizer Eduardo Peña and Lance Compa, a labor studies professor from Cornell University, who had just published a Human Rights Watch report on worker abuses in the nation's beef, pork, and poultry plants. I interviewed workers about their injuries on the job and a very busy workers' compensation lawyer. I was able to further dramatize some of the events inside the plant based on a 436-page ruling of a nine-month hearing into the labor issues, written by NLRB judge John West. I also interviewed Smithfield attorney Greg Robertson, who said the company was appealing the judge's decision because it was not an accurate reflection of Smithfield's labor practices — or what had happened.

But there was one major source I had not been able to track down: whistle-blower Sherri Buffkin. It was her testimony, as a former member of Smithfield management, that was key to the NLRB's case against the company. I had been able to quote her testimony before the Senate Committee on Health, Education, Labor and Pensions, but I knew she had much more to say.

In 2008, when I began working on my master of fine arts degree at Old Dominion University, the Smithfield Foods case was still ongoing, so I made it the focus of my thesis project. I knew I had to find the elusive Buffkin, who lived in rural North Carolina. On a Sunday afternoon in mid-February 2011, with the help of former union organizer Dan English, I found my way to her doorstep.

Buffkin didn't want to talk. She opened her front door a slit. "I've put all that behind me," she said. Her big dog Samo stuck his head into the slit, his booming voice making it hard to hear. I told her I knew how difficult it must have been for her to testify against her former employer. I told her that even if she didn't want to talk, it was an honor to meet someone who had taken on Smithfield Foods and won. She let the dog out, had a cigarette, then invited me in for "just a minute." Several hours later, I walked out and headed back to Norfolk. That began a series of telephone conversations, with what would best be described as a reluctant participant. She was one of many.

I knew working on this book would be tricky even before I started it. Almost all the court records were sealed from Smithfield's racketeering (RICO) lawsuit against the United Food and Commercial Workers International Union (UFCW), settled in October of 2008. Furthermore, both Smithfield Foods and the UFCW rigidly interpreted the judge's order to not disparage the other party as a de facto gag order; both cited the order in declining to participate in this book. Nonetheless, I made multiple attempts to reach significant figures independently. In most cases, they either did not respond or declined to comment. Neither the UFCW nor Smithfield accommodated my requests for interviews with current or former employees, including former CEO Joe Luter III and plant manager Jere Null, who plays a key role in this story.

Fortunately, two key people were willing to talk: Jasper Brown, the lead attorney for the NLRB against Smithfield Packing, and Gene Bruskin, former director of the UFCW's union organizing campaign in Tar Heel. There were multiple other interviews, and many attempts were made to reach those affected by the case, which became increasingly difficult as each year passed. Several died. Some I was never able to locate. Some, who had received settlements, were unwilling to talk.

I was also able to tell this story because the NLRB has kept every bit of documentation on the case, which it considers to be among its most important legal proceedings. They provided a searchable 7,910-page transcript of the NLRB hearing against Smithfield Packing, heard over a nine-month period between 1998 and 1999. With that, I had the complete testimony of about 130 witnesses, which meant I could piece together details from multiple witnesses to re-create events.

In addition, I got copies of two civil lawsuits from state courts to fill in other details. Later, through the Freedom of Information Act, the NLRB provided documents relating to Smithfield's appeals to its rulings, which were not available to the public. Through the courts, I got access to documents filed—prior to the settlement—in Smithfield's RICO lawsuit against the UFCW and multiple other parties.

Finally, I benefited from having access online to hundreds of articles reporters wrote for the *Fayetteville Observer*, which covered Smithfield Packing extensively through the years, as well as Charlie LeDuff's brilliant *New York Times* piece "At a Slaughterhouse, Some Things Never Die" and the

Pulitzer Prize–winning series "Boss Hog," published by the Raleigh *News & Observer.* And, while I was writing this book, Matthew Barr, of the University of North Carolina at Greensboro, finished his documentary *Union Time,* which provided live action of some events I had not attended.

Unfortunately, Smithfield chose to remain silent and not address the reasons it repeatedly and illegally kept the union out of its largest slaughterhouse for sixteen years. Then again, even before the 2008 gag order, when Smithfield executives and attorneys were free to speak, they were unable to make a convincing argument to support their position that they had done nothing wrong. Nor could they offer any compelling reason why—if they were truly innocent—a longtime judge for the NLRB would rule that virtually every one of their ninety-five witnesses had lied during that hearing, or why, even with additional evidence provided by the company, both a three-member appellate panel of the NLRB and a federal appeals court upheld that judge's ruling.

Smithfield finally agreed to comply with the NLRB's rulings requiring extraordinary policing of the final union vote. Still, company executives have never credibly explained—even on a business level—why it made sense to spend millions of dollars to fight to the bitter end to keep out a union they publicly stated they worked well with in other major plants. Perhaps one day, former Smithfield chairman Joe Luter III will explain.

# HOG WILD

On September 26, 2013, China's largest pork processor, Shuanghui International Holdings Ltd., purchased Smithfield Foods, the world's largest pork processor and producer. The $7.1 billion deal was the biggest takeover of a U.S. company by a Chinese firm. Put simply, the Chinese eat a lot of pork—85.3 pounds per person in 2012, compared to the 59.3 pounds eaten by each American—and they need a reliable supply for their population of 1.3 billion, whose standard of living—and ability to buy meat—is rapidly rising.[1]

Yet Smithfield still constitutes a tiny portion of today's supply chain. Although the company slaughters about a quarter of the hogs in the United States, it accounts for only 3 percent of China's slaughtering capacity, and China produced more than half the world's total pork in 2012.

Still, Smithfield's size and history make it worthy of scrutiny because it illuminates the high cost of putting pork on the table. One reason the Chinese bought Smithfield was to remake its own hog-slaughtering industry in the U.S. company's image, achieving the same money-saving, profit-generating efficiencies. But Smithfield's model for industrial meat production causes tremendous hardships for its workers, as well as polluting the water and air and inflicting unnecessary pain on the animals. Often it is the taxpayer who must pay for remedies, after the damage is done.

Slaughtering animals and processing them into food has always been a messy and unpleasant task, and exposés of the abusive working conditions in industrial meatpacking date back at least to Upton Sinclair's *The Jungle*, published in 1906, a book often credited with catalyzing the creation of the U.S. Pure Food and Drug Act that same year, which launched the federal government's regulation of Americans' food supply and pharmaceuticals. But remedying the low pay and dangers typical of mass butchery was left up to the workers themselves through unionization and collective bargaining, eventually protected by federal law through a series of laws passed during the first part of the twentieth century, culminating in the National Labor Relations Act passed in 1935 as part of the second wave of Roosevelt's New Deal.

The purpose of the act was to level the playing field so that businesses and workers could fairly negotiate the nature of their relationships in a win-win contractual agreement. Initially, it seemed to work. Through the early 1980s, meatpacking workers enjoyed wages about 14 to 18 percent (nearly $20 an hour adjusted for inflation) above the average for manufacturing employees, thanks to the United Food and Commercial Workers International Union (UFCW).[2]

But that quickly changed, and Smithfield Foods was in the vanguard of companies transforming the meatpacking industry to lower costs and make meat more affordable for ever larger numbers of consumers, first in the United States and then around the world. Unfortunately, this transformation has come at a high price for workers and the communities where hogs are raised and slaughtered. Americans are still paying this price today, despite decades of protest and dissent and imploring national lawmakers to stiffen or enforce labor and environmental law.[3]

The placement of a 973,000-square-foot plant in one of the poorest counties in North Carolina and the ripple effect throughout every layer of life surrounding it provides a window into meatpacking across the United States. It is a microcosm of fractured labor law and enforcement, where today in the United States more than 20,000 workers each year — up from just over 6,000 in the late 1960s — suffer illegal retaliation for exercising their federal rights to negotiate the conditions of their workplace without fear of losing their jobs.[4] Where 75 percent of companies involved in a union campaign hire anti-union consulting firms — called union busters — a booming $4 billion industry.[5] Where federal law enforcement of the NLRA is so weak and so slow that corporations simply factor the small penalties into their cost of doing business, at the expense of worker safety and wages. Where corporations can appeal federal judgments against them for breaking federal labor law, extending enforcement for years — sometimes more than sixteen years, long after workers would want their jobs back, long after they've found other work. To add insult to injury, any wages earned by workers during those years are deducted from monetary penalties paid by the company that illegally fired them. Thus, ironically, the federal government penalizes fired workers for bootstrapping and providing for their families, while the wheels of justice slowly grind up their idealistic belief in a land for, of, and by the people.

It also represents the meatpacking industry's national trend to move slaughterhouses away from unionized cities into poor rural areas in "right-to-work states" that discourage unions, as well as the common practice of using massive profits to grease the wheels of southern political will to take advantage of the poor. And it provides insight into how corporations have taken advantage of the influx of often undocumented workers—easily intimidated with threats of deportation—and pitted poor blacks against desperate immigrants to dilute worker solidarity theoretically encouraged by national labor law.

Smithfield Foods placed its mothership plant in southeastern North Carolina, where the company leveraged massive, well-oiled anti-union marketing on disorganized workers, picking off pro-union leaders one by one and using them as examples of what would happen to other workers if they tried to organize or even spoke up in favor of unionization.

This is, of course, illegal. But the loopholes under current weak—and unenforced—labor law make it possible for companies to get away with illegal firings, intimidation and even assault, with impunity for many years. "The consequences for a company that crosses the line are not that severe," said Wilma Liebman, chairman of the National Labor Relations Board from 2009 to 2011 and a board member from 1997 to 2009 as both a Republican and Democratic appointee. "They pay a little back pay, years down the road. For many companies, fines and paying back pay is just the cost of doing business."[6]

When passed in 1935, the National Labor Relations Act—the principal federal statute governing private-sector labor relations excluding the railway and airline industries—directly addressed the inequity of bargaining power between employees and employers and contributed to the "widely shared prosperity that prevailed after the Second World War."[7] This reversed the inequalities that existed at the turn of the century during the Gilded Age, when in 1915, 2 percent of the population owned 60 percent of the wealth and 65 percent owned just 5 percent of the wealth.[8] "In America in particular the share of national income going to the top one percent has followed a great U-shaped arc. Before World War I the one percent received around a fifth of total income in both Britain and the United States. By 1950 that share had been cut by more than half. But since 1980 the one percent has seen its income share surge again — and in the United States it's back to what it was a century ago."[9]

Today, the richest 1 percent own more wealth than the bottom 90 percent.[10] Some directly blame labor law, in dire need of reform, for actually "contributing to the demise of the unions it was initially enacted to protect."[11] And some directly link the demise of the unions with the demise of the middle class. That the middle class has been hollowed out is not a coincidence, according to economist Paul Krugman, who writes that "in the 1970s, corporate America, which had previously had a largely cooperative relationship with unions, in effect declared war on organized labor."[12]

Since 1971, the percentage of the population that is middle class has steadily declined from 61 percent in 1971 to 50 percent in 2015. Meanwhile, the percentage of the upper middle and upper classes has risen from 14 to 21 percent. The percentage of highest-income households more than doubled, from 4 to 9 percent. The shift away from the middle reveals a polarization in the economy.[13] After 1980, only the top 1 percent saw their incomes rise.[14]

In 1980, the National Labor Relations Board — charged with enforcing labor law in the private sector — had 44,063 cases. By 2014, that had dropped by more than 50 percent to 21,394. At the same time, union membership in the private sector dropped from 21 percent in 1980 to 6.7 percent in 2013, or by 69 percent.[15] In North Carolina, just 1.9 percent of workers were members of unions in 2015, down from 3 percent in 2013. Of states with lower than 5 percent membership, seven of the eleven are in the South.[16]

One reason for the drop in NLRB cases is that unions have simply lost faith in the NLRB, due to its dysfunctional protracted processes, delays, and weak remedies and the politicization of its board and decision-making.[17] The NLRB — with its five members nominated by the president and confirmed by the Senate — resolves disputes between unions and employers, conducts secret-ballot elections if workers show that most want a union, and adjudicates charges of unfair labor practices by employers and unions.

As Liebman has noted in several articles as early as 1983, citing others, scholars were calling labor law "an elegant tombstone for a dying institution."[18] By 2012, an NYU professor referred to its "ossification." On the occasion of the NLRA's seventy-fifth birthday, Harvard labor economist Richard Freeman declared that in 2010 "the law no longer fits American economic reality and has become an anachronism irrelevant for most workers and firms."[19]

While this was mulled by academics and progress was stymied in Congress, the result of inaction was stunning and swift in 2016, when workers — particularly white workers in Middle America — revolted, electing President Donald Trump, who promised to bring back the jobs lost through international trade agreements and union dissolution. He even met with union representatives on his first day in office. For the first time during a national election, the unions were unable to deliver votes to the Democrats, as disgruntled, out-of-work voters protested rust belts marked by abandoned factories.[20]

Pennsylvania, Florida, Ohio, Wisconsin, Iowa, and Michigan all turned red. "We underestimated the amount of anger and frustration among working people and especially white workers, both male and female, about their economic status," said Lee Saunders, president of the American Federation of State, County and Municipal Employees and chairman of the AFL-CIO's political committee.[21]

The roots of the discontent, former NLRB chairwoman Liebman believes, lie in the "erosion of middle-class lifestyles among average Americans and the loss of the expectation that each generation would live better than its parents. There is deep resentment about persistent long-term unemployment, the loss of jobs to foreign competitors and off-shoring, the outsourcing of entire classes of employment through digital technologies, rising inequality of incomes, corporate greed and public bailouts of irresponsible banks and international trade agreements biased in favor of big business at the expense of American workers."[22]

While Donald Trump's presidential candidacy exploited this "anti-regulatory fervor," Liebman believes that "no progressive legislation will come out at the federal level" for the next four years; what remedies emerge will come at the state level. In Republican circles, says Liebman, the NLRB and labor law reform are "toxic," particularly in the South.[23] And the NLRB under Mr. Trump will no doubt overturn numerous union-friendly moves by the Obama board, among them ones speeding up unionization elections.[24] "They will certainly overturn a lot of the Obama decisions," Liebman said.

"But, even though elections were speeded up, the union win rate has not improved, so they may not bother with overturning those rules."[25] Still, according to Liebman, Trump's Supreme Court pick will likely be bad for labor as well.

Today, just 11.1 percent of American workers belong to unions, half the level when Ronald Reagan became president and down from 35 percent in the 1950s. Jacob S. Hacker, a political science professor at Yale, said the shrunken movement, which represents just 6.7 percent of private-sector workers, faces "an existential crisis."[26] "There's an irony here," he said. "Unions are probably the most consistent voice for the broad middle class of any organization today, yet the voice of the middle class was seen as an important part of Donald Trump's victory. The further decline of labor is going to hurt many members of the middle class."[27]

Wages have stagnated for more than thirty years, while manufacturing has declined. And the right-to-work movement, which discourages workers from unionizing and weakens the strength of unions where they exist, has spread from the South — with its long antipathy to unions — to the rust belts and beyond. In February 2016, Republicans in West Virginia overrode the governor's veto to became the twenty-sixth state to adopt right-to-work laws. And there is a movement afoot to make right-to-work a federal law.

Still, Liebman points out that labor unions existed before labor laws were passed to protect workers rights and to end the often state-sanctioned, sometimes violent, bloody shutdowns of strikes. "Labor law is at its heart a human rights law," she said.[28] She also points to state-enacted increases in minimum wages. Stronger labor laws will follow from a stronger labor movement, not the other way around, she predicted. "Worker restiveness is widespread, genuine, and cannot be ignored, underscored by the unexpected and lasting appeal of both Donald Trump and Bernie Sanders. Reenergized worker power can catalyze political will."[29]

But worker power is greatly diminished. In the face of eclipse, unions have "shelved the strike weapon," according to historian Nelson Lichtenstein. "In 1999 there were only 35 strikes involving more than 100,000 workers; 25 years before, there had been ten times as many."[30] One reason is failed labor law today, especially in the South, where increasingly corporations are wooed with promises of factories free from worker demands for fair wages, which fails to promptly and fully support fired workers — especially meatpackers—who try to gain control over line speeds, safety issues, bathroom access, clean water, and a living wage.

"Smithfield was the quintessential example of the failure of the National Labor Relations Board," said one union leader. "When they beat you up, call you a nigger, and stand there with guns and you still can't get the relief you need, that's failure."[31]

## CHAPTER 1

# Joe Luter and Smithfield Foods

W hen gourmands think of the famously sought-after Smithfield Ham, they think of biscuits with salty slices of pink pork, sliced razor thin. Indeed, to bear the name Smithfield Ham, the cut of meat must be processed within the city limits of the quaint historical village of Smithfield in Virginia that dates back to the seventeenth century. But most of the hams produced by Smithfield—note the lowercase "h" here—come out of its sprawling slaughterhouse in Tar Heel, North Carolina, about 230 miles south of Smithfield Food's large, brick corporate headquarters on the banks of the Pagan River in the village of Smithfield.

From Smithfield, it's about a four-hour drive to Fayetteville down I-95, then onto Highway 87. There a steady stream of hog trucks pull into the silvery pipe-ridden industrial complex that looms up out of a sprawling, poverty-stricken flatland just northwest of the tiny town of Tar Heel. This is Smithfield Packing, long the dark underbelly of Smithfield Foods. There, starting in 1992, thousands of workers began to emerge from the bleak economic landscape of southeastern North Carolina to receive a starting hourly wage of $8.10 to $8.60, roughly twice the minimum wage of $4.25.

Today Bladen County, where the plant is located, still has one of the highest rates of unemployment in the state, at 6.9 percent, and more than one in four live below the poverty level (27.4 percent). Neighboring Robeson County, where many workers live, has a 32 percent poverty rate and unemployment of 7.9 percent.[1]

Every day in Tar Heel, in a 973,000-square-foot slaughterhouse, workers, mostly minorities—Latinos, blacks, Native Americans—gas, bleed, and disassemble up to 32,000 hogs a day. That's 16,000 hogs per eight-hour shift, 2,000 per hour, 33 hogs every minute, 1 every two seconds. It's punishing, mind-numbing work that leaves workers' muscles burning and hands cramped and tingling from thousands of cutting motions a day. It's bloody, smelly work with extremes in temperature as hogs are scalded and then chilled.

Meatpacking is the most dangerous manufacturing job in America today, with two and a half times the average injury and illness rate. Serious injuries requiring work restrictions or days off are more than three times higher than for U.S. industry as a whole. As many as 69 percent of injuries are never reported, according to the Bureau of Labor Statistics.[2] Even in a unionized shop — where stewards monitor line speeds and attend to workers' needs and injuries — workers risk debilitating hand and repetitive-motion injuries, gashes, amputations, and even death.

But when it opened, Smithfield Packing — a subsidiary of Smithfield Foods — wasn't unionized. At the time, it was the only major Smithfield packinghouse that wasn't. And owner and CEO Joe Luter III clearly wanted to keep it that way. One big reason was to make sure there was no chance of a production shutdown. "If we had an extended strike in that plant," Luter said, "we would have hogs backing up throughout North Carolina or we would have to put hogs on trucks and ship them to the Midwest."[3]

Smithfield Foods started in 1936 as a small family slaughterhouse in Smithfield, Virginia, then grew, through aggressive takeovers, to a Fortune 500 company with a record operating profit of near 20 percent gains in the first half of 2016 after earning $14.4 billion in 2015, selling its products through brands such as Armour, Morrell, Gwaltney, Nathan's, Farmland, and Ekrich.[4] Most of its operations are based in the United States, with a few in Mexico, Romania, Poland, and the United Kingdom.

In 1992, the corporation spent $80 million to move the center of its slaughtering operations to Tar Heel, North Carolina. Today, Smithfield owns most of the hogs in a state that has more hogs than people, second only to Iowa. Altogether, nationwide, Smithfield runs nearly forty hog slaughterhouses and meat-packaging houses, mostly in the Midwest and southeastern United States. It runs nine in eastern Europe, eight of which are in Poland. Smithfield also dominates the national hog-raising arena, with company-owned farms primarily in North Carolina but also in Missouri, Utah, Colorado, Virginia, Oklahoma, Illinois, Texas, and South Carolina. It's difficult to fully assess its power and control because of individual contracts with untold numbers of independent farmers in North Carolina, South Dakota, Colorado, Iowa, Missouri, Oklahoma, Pennsylvania, and Virginia.[5]

Joseph "Joe" Williamson Luter III was born on July 17, 1939, three years after his father and grandfather—a meatpacker born in Ivor in Southampton County in 1879—opened Luter Packing Company, later re-named Smithfield Packing Company, in Smithfield, Virginia.[6] Like his father, Joe Luter III grew up in Smithfield, which sits on the winding Pagan River in Isle of Wight County, just south of Jamestown. As early as the 1700s, the town had become known for Isle of Wight Bacon and Smithfield Ham.[7] The first extant record of the industry in America is an invoice dated April 30, 1779, from an early Smithfield packinghouse. By the 1800s, Smithfield residents who prospered from trade—including pork and peanuts—had built elaborate Victorian homes, "their ostentatious elegance visibly evident," with turrets, towers, stained-glass windows, and steamboat-style Gothic trimming around the original colonial cottages at the center of town.[8] In 1926, it became illegal for any pork that had not been processed within city limits to bear the name "Smithfield Ham," sometimes called "the aristocrat of the Virginia table."[9]

Joe Luter grew up in the heart of this southern charm. He and his sisters lived at the top of the hill, about three hundred yards from where company headquarters sits today on the banks of the Pagan River. Life was quiet and conservative, he said. "My father never left the house unless he had a hat on, along with a jacket and tie. My mother never left the house without wearing short heels and a dress."[10]

His father and grandfather both learned the trade at meatpacker P. D. Gwaltney, Jr. & Co. until they decided to go out on their own, providing the sweat equity in a new company, Smithfield Packing Company. In 1936—the same year P. D. Gwaltney Jr. died—they began curing Smithfield hams and selling them to mom-and-pop stores in nearby towns. Ten years later, they built a slaughterhouse on Highway 10, expanding until they were slaughtering about 3,500 hogs a day. By 1959, they employed 650 workers. Luter Jr. was also secretary-treasurer of Luter Packing Company in Laurinburg, North Carolina, less than fifty miles from Tar Heel.

Luter Jr.—who had gone to work part-time in a Smithfield slaughterhouse at the age of 12—was a workaholic, his son said. "He was in the office six days a week and five nights a week."[11] In May of 1946, when Joe was about 7, the family must have watched closely as the United Packinghouse Workers of America (CIO) attempted to unionize competitor P. D. Gwaltney's 80 percent black workforce of about 114 workers. The union had scheduled an NLRB secret-ballot election on May 29. About half the town's 1,300 citizens were black and, according to the NLRB, it was the first attempt to organize a packinghouse in the town.

If Gwaltney was opposed, it let others do the dirty work. The publisher of the *Smithfield Times* ran anti-union ads, formed an anti-union committee, sent anti-union postcards to Gwaltney employees, and put together a petition opposing the CIO. The publisher called a meeting for May 27, attended, according to the NLRB, by three to four hundred people, the largest gathering ever held in the community. Attendees were equally white and black and included many Gwaltney employees. The principal speaker was Remmie L. Arnold, president of the Southern States Industrial Council, an association that opposed New Deal policies. Arnold warned of communism, of another "Reconstruction," and included threats, veiled as recounted incidents of Ku Klux Klan terrorism where "hooded figures galloped through the night striking terror to the hearts of Negroes. . . . Today occasional flaming crosses again light Southern skies as CIO and A.F. of L. move in. . . . Gentlemen, a current drive by the Communist supported CIO-PAC may raise the cry: 'The Klan Rides Again.' May God forbid this."[12]

In the days before the union vote, the county sheriff threatened to drive a union representative "out of town." And the police chief offered to give the representative "a salt water bath" and told the representative the city was "way behind in lynching around here. We haven't had a lynching in about 20 years."[13] On May 29, both the sheriff and the police chief positioned themselves prominently in front of the entrance of the warehouse where the vote was held. The union lost the vote 85 to 27.

The union filed an objection with the NLRB, and a four-day hearing was held that December in Smithfield. Two women testified they overheard conversations about "getting rid" of union representatives. The NLRB determined "that threats of violence were uttered" and set aside the results after determining that "a hostile and threatening" atmosphere including mentions of "flaming crosses" and the "Ku Klux Klan" may have made the employer's black employees think they might suffer physical violence if they voted for the union.

There was no reported evidence that Gwaltney owners or managers had been in collusion with the anti-union movement; nonetheless, the NLRB set aside the election because the atmosphere prohibited a free election.[14] Less than ten years later, on March 3, 1955, by a vote of 289 to 176, apparently without incident, workers at Smithfield Packing voted to unionize.[15]

Joe Luter was 14, and if he was not already working at his father's plant, he soon would be. He has apparently never spoken publicly about the unionization of the meatpacking industry that dominated his small hometown of Smithfield, but he must have been aware of worker sentiments as he moved from one job to the next inside the plant. As he told the NLRB, "I've loaded trucks. I've worked on the kill floor. I've worked in sliced bacon. I've worked all through the plant when I was in high school and college."[16]

But unlike most slaughterhouse workers, this young man had the opportunity to move on to less strenuous and dangerous work. He graduated from Wake Forest University with a degree in business administration, and when his father died unexpectedly in 1962 — he had diabetes and a heart condition and smoked three packs of cigarettes a day — Luter III stepped in. He bought out nonfamily owners and, at only 26, became president of the company.[17] His son, Joseph W. Luter IV, was born on February 26, 1965, joining older sister Laura. The youngest, Leigh, was born a year or two later. He wanted to be a family man, not a workaholic like his father; he said, "I'm not like that. I have other interests."[18]

At the time he took over, the plant was slaughtering about three thousand hogs a day; by the time he left in 1969, it was up to five thousand. In July of that year, he sold the controlling stakes of the company to a Washington, DC– based conglomerate, Liberty Equities, for $20 million in cash and debt notes. By January he had been fired.[19] The now very wealthy young man left hog slaughtering behind to start up a Shenandoah ski resort, Bryce Mountain, selling unbuilt lots from 1970 to 1975. "I had a beard and was wearing blue jeans and running a ski resort. I was perfectly happy," he said.[20] Luter sold 2,500 lots and built a golf course, a lake, a clubhouse, tennis courts, stables, condominiums, and townhouses. But he was restless.[21]

Liberty Equities, owned by the flamboyant C. Wyatt Dickerson Jr., who would become part of Washington's glitterati, was accused by the Securities and Exchange Commission (SEC) of falsifying earnings reports to pump up the price of its shares. The SEC halted trading on the stock, the chairman resigned, and new management ran Smithfield Packing into the ground in just a few years. In 1973, it lost $3.6 million, and another $8 million in 1974, when it had exhausted cash reserves and had debt payments due.[22] The conglomerate sold off most subsidiaries and changed its name to Smithfield Foods.[23]

In 1975, Luter returned from the mountains to salvage what he could of what was left of the two remaining units: Smithfield Packing and the Family Fish House chain. Taking advantage of the hog price cycle to acquire stock when the price of Smithfield shares was depressed because of high commodity prices, Luter bought the company back for ten cents on the dollar, and in one year turned the $8 million loss into a $393,000 profit.[24] In April of 1975, he became president and CEO, later joking about Liberty's inability to sell Christmas hams: "For a company like this to lose money in December, it's like Budweiser losing money in July."[25] Luter began a six-year business reorganization with $17 million in long-term debt, a net worth of less than $1 million, and stock valued at 50 cents a share. He slashed costs. In 1978, Luter sold the Family Fish House chain for $7.75 million, using the cash to purchase troubled meat-processing plants for less money than he would have needed to build new ones.[26] "I really didn't want to see the company that my father and my grandfather spent their lives in fail," Luter said.[27] In 1981, the company became the dominant meatpacking force on the East Coast when Luter bought out Gwaltney of Smithfield, its fiercest competitor. "We were talking one night, and Joseph was going on and on about Gwaltney," said Barbera Thornhill Luter, who was married to Luter in the late 1980s. "Finally, I said, 'If you can't beat 'em, why don't you buy 'em?'"[28]

The purchase, for $32.5 million, doubled the company's sales to $600 million a year, but the purchase caused "some assimilation problems because we had been major competitors for so many years." The bitter rivalry, which perhaps started when his father and grandfather left Gwaltney years before, never died. To manage animosity, Luter created separate subsidiaries with separate management, sales, and production staffs.[29]

The major recession in the Nixon-Ford era, which essentially ended the long post–World War II economic boom, created the perfect opportunity for a man like Luter.[30] Luter had an aggressive (enemies say ruthless) character.[31] He admits that he's "combative by nature," and has been described as uncompromising, opinionated, and vindictive. It has been said that he can't stand taking orders from the government and savors fighting back.[32] He likes to hunt grouse and waterfowl and has a taste for Ferraris—at one point owning a Ferrari Testarossa—and BMWs.[33] He has had estates in Smithfield,

Washington, DC, and Manhattan. There was a reason Luter became known as "boss hog."[34]

By 1982, Luter was divorced from the mother of his three children and had married Barbera Thornhill, a native of Raleigh and longtime resident of Washington, DC, where the couple lived and where she ran Impact Design, an interior design firm with multimillion dollar jobs with clients including the royal family of Saudi Arabia. He had three main hobbies, she said: "hunting, hunting and hunting." She also said he was ultracompetitive at cards and backgammon. "If you play either of those games for fun, don't play with him."

Despite his ability to run his company far from his factories or headquarters, as well as his determination to avoid his father's long hours, his diversions were an illusion, he admitted. "My mind never leaves the business. I wake up every morning thinking about it, and I usually go to bed at night thinking about it, and I guess 90 percent of my conversation outside the office is centered around business."[35] A visionary strategist and entrepreneur who did not like to get involved in minutiae or micromanaging, he always liked building a company better than running it.[36] And the 1980s were a good time to be reimagining meatpacking.

––––––––––

Meatpacking began to transform itself in the 1960s, looking for efficiency and higher profit margins. By shutting down old plants and building new ones in rural areas near cattle feedlots and swine-raising production sites, companies could make inroads both in labor costs, by busting old unions and preventing unionization at new plants, where organizing was difficult, and in transportation costs, by switching from rail to trucking.

The number of packinghouse workers in urban areas fell by more than 50,000 between 1963 and 1984, while the rural workforce doubled from 25 to 50 percent of the national meatpackers. Although companies did reduce transport costs by moving closer to the animals, the most significant gain came from reduced labor costs. In 1952, one man–hour of labor produced 51.4 pounds of dressed meat; by 1977, that had tripled to 154.6 pounds.[37]

By the late 1980s, the big meatpackers had virtually broken the back of organized labor. In a few decades, slaughterhouse workers went from being some of the highest paid to some of the lowest. As union protection faltered, workplace injuries soared. It was a bitter defeat, coming just decades after slaughterhouse workers successfully fought for their rights.

Nearly a century earlier, in 1897, the Amalgamated Meat Cutters and Butcher Workmen of North America became the "first national organization dedicated to bringing up working standards of the meat industry through unionization."[38] It represented workers at the "Big Four," including Armour, Swift, Wilson, and Cudahy, which had stockyard plants in Chicago, Kansas City, and Omaha. Between 1900 and 1904, workers gained seniority rights, and set wages.

In 1904, employers reduced wages and tens of thousands of Amalgamated members walked off the job in a nationwide strike. Employers hired thousands of African Americans as strikebreakers, rightly gambling that the unions would not be able to organize across racial lines.[39]

The industry was indeed an amalgam of ethnic backgrounds. After the turn of the century, Poles replaced the Germans and Irish as most prominent, followed by Slovaks and Lithuanians. Blacks from the Deep South began immigrating after 1900, and Mexicans began arriving during the early 1920s. After passage of the National Industrial Recovery Act in 1933, new packinghouse unions were forming—including the Packinghouse Workers Organizing Committee (PWOC)—that succeeded in overcoming ethnic and racial antagonism, representing blacks, whites, and women. In 1943, the PWOC became the United Packinghouse Workers of America (UPWA). The blending of union support and civil rights objectives laid a historical foundation for worker unity across racial lines at Smithfield Packing in Tar Heel some fifty years later.

Throughout World War II and into the 1950s, packinghouse wages rose to 20 percent above average manufacturing wages. Into the 1960s, at some plants the union still "had a fairly firm handle on chain speeds and proper crewing," one Morell Sioux City worker said then. It was at "a pace that you could handle. You could do the work and get it done without killing yourself." The union steward had as much authority as the foreman did, the worker said.[40]

Iowa Beef Processers (IBP) was the first to break free, revolutionizing by using automation and unskilled rural labor. Founder A. D. Anderson told *Newsweek* in 1965: "We wanted to be able to take boys right off the farm and we've done it."[41] Workers initially fought back. In 1969, striking workers temporarily closed four plants in Iowa and gained a salary increase of

20 cents an hour, still far below previous union wages. The pork industry quickly followed IBP's lead. Between 1956 and 1965 Armour closed twenty-one major plants that employed 14,000 workers, replacing them with a dozen modern plants employing fewer than two thousand workers. In a decade, meatpacking union membership dropped precipitously from 59,550 in 1953 to 26,600 in 1964. An aggressive recovery in organizing at new plants was no longer possible.

The United Packinghouse Workers of America (UPWA) merged with Amalgamated in 1968, preserving control of local packinghouse unions in plants owned by national firms for a time. But in 1979, when Amalgamated merged with the Retail Clerks International Union (RCIU) to form the United Food and Commercial Workers (UFCW), the decline of union control over work conditions in slaughterhouses accelerated.

Infighting in the newly formed union—too thinly spread over diverse workplace cultures—helped further weaken an already lost cause. Less than 10 percent of the membership were packinghouse workers. Former RCIU officials were not accustomed to national industrial horse-trading or rough in-your-face negotiations. The clash was resounding as RCIU holdouts resisted efforts of packinghouse workers to fight back against the decreasing wages, increased line speeds, and safety hazards that came with automation.

Even in the Reagan era of fired air traffic controllers and union concessions to automakers, the decline of union power in meatpacking was dramatic. Reagan's political appointments to the NLRB removed any federal remedy for bullying of workers. The UFCW was simply ill-prepared for the challenge to union strength in the 1980s.[42] Aside from Richard Nixon's setting up the Occupational Safety and Health Administration (OSHA) in 1970, no major worker-oriented social legislation has been passed since the 1930s, and the limitations in enforcement of national labor law remain legion.[43]

The new "Big Three" meatpackers—IBP; ConAgra in Omaha, Nebraska; and Cargill in Wichita, Kansas—were shrewd. Slaughterhouse workers weren't the only place they were looking for savings. They also were cutting out the middlemen by procuring animals directly from independent producers and selling packaged meat directly to grocery chains. After World War II, many packers had mechanical stunners, conveyors, overhead chains, forklift tractors, sausage-making machines, vacuum packaging, motorized knives, and saws. But the Big Three systematized and integrated these technologies.

Most revolutionary, the new Big Three placed cutting and packaging operations in the same plants where the animals were killed, dispensing with the need for transportation of meat to other locations for packaging and eliminating the need for skilled knife work, thereby reducing labor costs on the back end, traditionally paid to highly skilled butchers in grocery stores. Then they viciously undercut each other in large sales to supermarket chains and other outlets.[44]

By the end of the 1980s, the power of unions to negotiate a single deal governing wages for chains of meatpackers had virtually ceased to exist. The UFCW was forced to grant repeated concessions as "wages and benefits fell precipitously . . . and the dominant firms in meatpacking became predominantly non-union operations."[45] Plant closings and the shift to non-union work forces "resulted in rapid declines in the number of workers covered by the UFCW master agreements," and between 1976 and 1983, meatpacker membership fell by a third to 30,000. By 1990, average hourly earnings for packinghouse workers were 20 percent below the average in manufacturing.[46]

What union power remained was severely compromised and ineffective. Coordinated national bargaining had disappeared. Regional—not national—directors were negotiating with packing firms. Still, there were feisty pockets, especially in Wisconsin and Minnesota. Smithfield purchased Patrick Cudahy in Wisconsin in 1984, but workers went on strike when Luter reduced wages to make the company profitable.[47] "He was especially concerned about the older employees who had been there many years," Barbera Thornhill recalled. "He really wanted to save the company and the employees too, but out there, it's union, union, union—even if it cuts your family's throat."[48]

Luter wasn't used to labor problems at his Virginia and North Carolina plants. The most serious had been a two-week standoff at one plant in 1969. "There's certainly more militancy in Milwaukee," Luter said. "In that part of the country, the workers consider themselves union members first and employees second. They forget who writes their paychecks. I don't think that is true in other parts of the country."[49]

## CHAPTER 2

# Cheap Labor Built on a Legacy of Slavery

To expand a growing empire with a centralized slaughterhouse on the East Coast, the astute capitalist Joe Luter needed four things: a place where he could partner with politicians; a place with minimal or easily subverted environmental legal restrictions to accommodate both a massive plant and the enormous anticipated increase in the number of hogs raised near the plant; legal loopholes allowing a near-monopolistic control of the supply of pigs; and low labor costs in a non-union plant. A right-to-work state with a history of oppressing workers, maybe even a legacy of slavery, would be just the place. North Carolina—with the lowest union membership in the country—fit the bill on all counts.[1]

There was only one problem. North Carolina had spewed forth a spitting, scratching tomcat of a girl, not unlike many other girls at that time trying to claw their way to a life better than their mama's. But this one girl just happened to grow up to work for Joe Luter. And she was a beauty. Strong, willful, smart, and blonde. And not afraid to curse, pull a gun, or beat the living daylights out of anyone who crossed her.

Sherri Wright Buffkin was born and raised in Bladen County, North Carolina, where peanut, tobacco, and cotton fields sprawled across a poverty-stricken flatland, near small towns where closing cotton mills had left behind desperation and welfare. Every day, as a child and teenager, Buffkin looked smack into the face of that poverty. The country roads carried her past families who couldn't find jobs and barely had enough to eat. Joe Luter was 30 when Sherri Buffkin was born on April 6, 1969. While Luter's father had been a workaholic, perhaps cold and distant, Buffkin's was an alcoholic with a temper who beat her mother. Luter's father could provide his children job opportunities when they came of age. Buffkin's father could not.

During the 1980s, while Sherri Buffkin worried herself through puberty, Luter—now a wealthy businessman—saw something shimmering above a flat swath of land just southwest of the Cape Fear River. It looked like gold. Buffkin had grown up on that river, sliding or climbing down rope ladders and chicken wire her daddy had strung up on its steep banks so the kids could get to the rope swing and drop into the water. Driving from home in Elizabethtown to Fayetteville, they'd pass that parcel of land and never pay it no mind. It was just 830 acres of nothing, right past the tiny town of Tar Heel on North Carolina Highway 87, a two-lane highway that wandered up to I-95, the largest north-south trucking corridor on the East Coast.

"My mama worked for minimum wage. If I got a shirt and a pair of shoes before school started, I was doing good; I was above all the other kids," Buffkin recalled.[2] There wasn't much to do but sneak out her bedroom window—or leave it open—so Buffkin got thrown out of her mama's house at 16 and married Davie, the only boy she'd ever been with. Her daughter Nicole was born March 12, 1991.

There was little work. Just an endless horizon. Of course, way back then, that Virginia businessman—with his obsession for building a pork dynasty—had never heard of Sherri Buffkin. But one day, he would. One day, he would know all about that poor little white girl from Bladen County.

———————————

A lot of people in southeastern North Carolina were poor, the result of the unforgiving land, sparsely populated even in precolonial times when Native Americans simply passed through using trading and warrior paths that ran parallel to the Cape Fear River that splits Bladen County. But the river didn't empty into a deepwater port like Charleston to the south and Norfolk to the north. It passed through Wilmington before dissipating into shallow waters inside a small gap between Oak Island and Bald Head Island, where it finally met the Atlantic Ocean. There, "shallow draft vessels made their way through barrier-island inlets into the shallow water of inland sounds."[3] Vast swamps and numerous streams made building roads and railways difficult. Nevertheless, Wilmington was one of two towns in antebellum North Carolina with a population greater than five thousand.[4]

Limited plantation work in southeastern North Carolina south of the river—where the soil and climate made it unprofitable to grow tobacco and cotton on as large a scale as plantations in Virginia—meant there were fewer slaves, concentrated in the "turpentine orchards" rather than the cotton fields.[5] Naval stores—products of the gum of the longleaf pine trees abundant in southeastern North Carolina—were North Carolina's leading industry. By the 1770s, the colony was responsible for 70 percent of the tar exported from North America and 50 percent of the turpentine.[6] But the boom would last little more than a century. Tapping the pine trees weakened the trees, making them susceptible to disease. Free-ranging hogs ate pine cones and rooted seedlings, interrupting the natural reforestation process. As a result, little of the longleaf pine forest remains.

Still, in 1790 North Carolina had the fourth largest number of slaves in the country. In 1860, a year before the start of the Civil War, one-third of the population were slaves (331,059 out of a total population of 992,622). Another 30,463 free people of color also resided there. Statewide, just over a quarter (27.7 percent) of families owned slaves, but in Bladen County nearly half of all families owned fellow human beings.[7]

When slavery was abolished, most stayed where they were, having little choice but to substitute tenant farming and sharecropping for slavery, often drawn into a spiral of debt, dependent on a single money crop—most often small patches of cotton or tobacco. The average size of farms began to decline, but farming, and then cotton mills, sustained the local economy in the late nineteenth and early twentieth centuries.

When the Civil War had ended, violence in the South did not dissipate. Whites remained entrenched in the lifestyles their slaves had built for them, and North Carolina was not immune. Lynchings became "an act of terror meant to spread fear among blacks" and maintain "white supremacy in economic, social and political spheres." They increased when the former slaves began to register to vote, start businesses, and run for public office. About five thousand blacks were lynched from 1880 to 1955, but North Carolina certainly wasn't the worst among the states, ranking thirteenth in the nation.[8]

Blacks weren't allowed to work in the powerful textile industry that dominated southern culture from the early 1800s and remained a "rigidly segregated institution" until the mid-1960s.[9] Massive union drives in the 1940s—including "Operation Dixie" in 1946—failed to organize textiles and "left Southern Dixiecrats and the system of white supremacy with complete social, political and economic hegemony intact in much of the South."[10]

In 1947, the passage of the Taft-Hartley Act allowed North Carolina to enact right-to-work laws, which permit workers to refuse to join a union or pay dues even if they are represented in their workplace. This robs the union of one of its most powerful tools: the ability to rally the work crew to walk off the job in protest over low wages or inhumane working conditions. The Taft-Hartley Act also sought to control communist infiltrations into the unions. The "Red Scare," with its second wave under Senator Joseph McCarthy, was "interwoven with the racism and hostility to outsiders endemic to Southern small town life." In turn, it overlapped with anti-union sentiment in "law and order leagues" led by business and politicians, which overlapped with the Ku Klux Klan and others.[11] These groups lumped civil rights, union collective action, and communism together, conveniently leaving white racial supremacy firmly entrenched. After World War II, the powerful textile industry "paired political power with race-baiting and other illegal practices to break the few successful unions in the region or keep the unions from taking hold."[12] Although repeatedly sanctioned by the NLRB, powerful textile companies chose to pay millions in fines and compensation to illegally fired workers rather than accept a unionized work force.[13] It was a complicated world indeed, where even pro-labor organizers could be "virulently racist." The racial gap proved unbridgeable. "There were three things that were in the same category and the same danger," said Thomas Knight, then secretary-treasurer of the Mississippi chapter of the AFL-CIO union. "A union representative, a black and a mad dog. A person with a gun in his hands would look at the three just about alike."[14]

Gorrell Pierce, former grand dragon of the Federated Knights of the Ku Klux Klan, was much more pointed about the links between trade unionism, communism, and matters of race in the documentary film *Resurgence*:

> Times are getting bad. What are we going to do, when Ford folds, when GM folds? How we going to bail them out? We're not. How are we going to bail out the textile worker in this state, the furniture worker? There IS no way. We're all going to be unemployed one of these days, people. When you're unemployed, what are you going to do then? I'll tell you what I'm gonna do. You're going to see me going down the street — After the son of gun that caused it: Africans, blacks, coloreds, niggers, negroes, or

whatever you call 'em. I call them black . . . (from the crowd: How about porch monkeys?) that's OK, porch monkeys is OK. You can find 'em in almost any union. And probably some of you here belong to a union. And unions have been good organization one time. But they've done gotten so big and powerful they're plumb out of the hands. Even the union employ-ees. Now the Teamsters is trying to do away with shop stewards—they gonna appoint 'em, you don't elect 'em no more. And I know why they wanna do that. They are going to elect you the finest little Communist, get in there and work in your union, he's gonna get you and your black brother all hugged up, and you're gonna talk about equality and fair pay, fair pay, and you're going to be going up and down the street marching together. And the next you know you're a Communist. And that's where they start. And they mean to physically overthrow this country! The Ku Klux Klan has never overthrew the government. We overthrew the Reconstruction government and replaced it with what it was supposed to be. And that's what we are here today to do.[15]

Finally, by the mid-1960s and early 1970s, after every major textile com-pany operating in the South was sued for alleged racial discrimination, blacks were allowed to work in the cotton mills.[16] By the 1970s, North Carolina led the nation in textile manufacturing, with over 400,000 jobs. But dreams of a unionized workforce were never realized, and by the turn of the century textile manufacturing had moved overseas or automated, leaving closed mills and blighted towns behind. By the end of the twentieth century, blacks in Bladen County were resigned to high levels of poverty and unemployment, with deeply rooted racial divides compounded by decades of welfare and desperation.[17]

## CHAPTER 3

# Lots of Pigs, Lots of Poop, Lots of Politics, Lots of Pollution

For Joe Luter, cheap labor was obviously a given. But a massive slaughterhouse needs something else. It needs pigs, lots and lots of pigs. Luter borrowed an idea from the poultry industry called vertical integration—owning the animals as well as the slaughterhouse—to smooth out the highs and lows of the market. By the end of 1987, he intended to have 10,000 sows in large pig farms in Isle of Wight County next to Smithfield. With one-third of the 17,000 hogs slaughtered daily by Smithfield coming from the Midwest, the saving in shipping, lost life, and weight loss would be tremendous. "We're the first pork packer in the United States to move aggressively in this direction," Luter said. "I think this will ensure the success of Smithfield Foods in the coming years."[1] Increased slaughter capacity combined with increased hog production was the key to success in the 1990s. "Quite frankly . . . the main reason I put the plant in North Carolina was that virtually 95 percent of the hogs (we slaughtered) were being raised in North Carolina," he said when he was chairman and CEO of Smithfield Foods.[2] "It saved me millions of dollars a year in transportation by putting the plant on top of the hogs."

North Carolina had a lot of pigs, but it didn't have enough pigs. A voracious slaughterhouse would need more and more and more pigs. Luter worked all that out. When construction on the new plant began in 1991, Smithfield Foods was buying 35 percent of their hogs from the Midwest, with increasing competition from IBP, which had entered the pork-packing industry. "We looked at our competitive advantages and disadvantages and came to the conclusion that our one major disadvantage was our sourcing of livestock," said Luter.

He was already working with the nation's fifth-largest pork producer, Carroll's Foods, headquartered in North Carolina, which in 1986 agreed to sell all its hogs to Smithfield Foods and purchase 13 percent of Smithfield's stock. Next, Luter joined forces with powerful politicians and leading state hog producers, formulating a business plan that would cause North Carolina's hog population to explode by 500 percent between 1992 and 1998, from 3 million hogs to 10 million, making it the second-largest hog producer in the nation, after Iowa, where it remains today.[3]

Luter partnered with North Carolina senator Wendell H. Murphy.[4] Murphy Family Farms, founded in 1962 and headquartered in Rose Hill in Duplin County, became the nation's biggest hog producer between 1982 and 1992 while Senator Murphy served in the General Assembly working hard to pass laws to benefit his industry, as the Raleigh, North Carolina, newspaper *The News & Observer* reported in its 1995 Pulitzer Prize–winning series "Boss Hog."[5] Murphy was "a back-room wheeler-dealer" who both sponsored and voted for tax breaks for hog producers together with protection from tougher environmental regulations. One of Murphy's coups was co-sponsoring bills to prohibit individual counties from zoning against factory farms. He also made sure environmental regulations were either not enacted or not enforced.

Murphy made illegal campaign contributions of $100,000 to help eliminate sales taxes on the hog industry, but when the state's second-largest newspaper pointed that out, law enforcement declined to prosecute, saying the two-year statute of limitations had expired. The tax exemption has been worth tens of millions—if not billions—of dollars to the hog-raising industry.[6]

Members of North Carolina's General Assembly were allowed to make money off the bills they introduced, as long as they could say their financial interests didn't cloud their judgment. It's hard to say that Murphy's judgment was in any way clouded. When Murphy was elected in 1982, Duplin County—where his business was headquartered—was home to 172,000 hogs. By 1995, there were more than a million. Most belonged to Murphy Family Farms.

Luter expanded his alliance with Carroll Foods and Murphy Family Farms to include Prestage Farms and Goldsboro Hog Farms, altogether referred to as "The Circle." "The Bladen County facility . . . was built to allow North Carolina's hog industry to grow," Luter said. "The gamble paid off. We went from a small, regional packer to the largest pork packer in the world. . . . For the most part, we eliminated the traditional animosity between packers and producers. . . . We've worked together and we've all done extremely well, compared to our competitors."[7]

Smithfield certainly wasn't alone. By the mid-1990s, more than 80 percent of hogs in the country were raised on farms that had coordinated agreements with processing plants. These arrangements allowed owners to coordinate the construction of packinghouses to match the rapid expansion in hog production. They patterned this business strategy after Tyson Foods, which led the industry in vertical integration.[8] "What we did in the pork industry is what Perdue and Tyson did in the poultry business," said Luter. "Vertical integration gives you high quality, consistent products with consistent genetics. And the only way to do that is to control the process from the farm to the packing plant."[9]

From the 1960s through the 1990s, meat costs fell as it took less time to grow a bigger chicken on less feed. The trend was mirrored in the hog industry. Hidden within this boon for consumers were taxes to fund billions of dollars in farm subsidies, used primarily to grow corn for animal feed. In 1994, direct payments to farmers were $7.9 billion. That grew to $12.4 billion in 1998 and $21.5 billion in 1999. The result was massive overproduction, a market glut of corn, wheat, and soybeans. Prices plummeted, creating an economy where it was cheaper to buy corn than to grow it. Industrial hog producers saved $947 million a year in feed costs.[10] The results weren't as good for small farmers. Within two decades of the reengineering of the hog industry, 90 percent of all hog farms disappeared in the United States, and the independent hog industry was wiped out.[11] In one generation North Carolina's small family farm industry transformed into large factory farms. In 1986, there had been 15,000 hog farms; by 2006, only 2,300 remained.[12] Between 1983 and 1995, more than 16,000 hog farmers left the industry, about two-thirds of the 23,400 growers.[13] Luter showed little sympathy for the thousands of small farmers shut out of the business by the arrangements with the meatpackers. "Small farmers have been disappearing for 100 years," he said.[14] "If you want to protect the small farmer, you are going to do it on the back of the American consumer."[15]

While midwestern states debated and passed protections for family farmers, North Carolina's government promoted large-scale producers. Critics say the hog companies rode to wealth on the backs of the remaining farmers, who took on large amounts of debt to build huge barns with thousands of small cages to raise stock they didn't own. "It's sharecropping," said one contract farmer.[16] Environmental activist Robert F. Kennedy Jr. agrees, condemning Smithfield's business plan in North Carolina as unethical. Kennedy has long documented the cost of Smithfield Foods' way of doing business and personally fought—unsuccessfully—to keep it from expanding into eastern Europe. "They built a big hog plant, then contracted with Wendell Murphy to produce hogs in confinement," Kennedy said in an interview. "Murphy rewrote twenty-four laws in North Carolina to make it impossible for local or county officials to zone them out or sue them. . . . They get hog farmers to sign a contract, mortgage their property, get the loans, pay the insurance, pay the heat, get rid of the waste—Smithfield says we'll only own the pigs, you'll own the poop—and then we'll give you a one-year contract though it will take you twenty years to pay off your hog house. He's now an indentured servant or a sharecropper."[17]

In contract farming, hogs are typically raised in three phases. Piglets are born at a farrowing operation where sows live in gestation cages so small the pregnant hogs can't stand up or turn around. Feces drop through metal slats, exposing sows to high levels of ammonia and respiratory problems. Neurotic behavior from the hogs—smarter than dogs—includes obsessive biting at the bars.[18] The cages have been banned in Europe, and in 2007, Smithfield promised to convert their gestation cages to group housing. By early 2017, it had completed about 87 percent, though it apparently confines sows for several weeks at the beginning of their pregnancies.[19] Contract growers—which hold about 40 percent of Smithfield's sows in 2,400 farms—lag significantly, with a completion goal of 2022. Smithfield officially recommends that they convert, but it is not mandatory.[20]

After the piglets are weaned—from suckling their mothers through bars in cages—they are trucked to nursing operations where they are kept in thousands of small pens inside long barns. They will not see the light of day again until they are loaded onto trucks headed for finishing facilities, where they are fed high-calorie grain until they reach more than 250 pounds at about six months of age. When they can barely fit in their holding cages, they are loaded into trucks headed for Tar Heel.

Under this contracting system, farmers borrow large amounts of money to build barns, pits under the barns to temporarily hold waste, and waste lagoons, often as deep as twenty feet. It can cost as much as $1 million to start an operation. Smithfield Foods dictates the details of how the hogs are raised, fed, and immunized. But, as Kennedy pointed out, they don't tell their contract farmers what to do with all that poop.

By 1995, 7 million hogs were producing about 9.5 million tons of manure a year, stored in thousands of lagoons, then sprayed onto surrounding croplands, saturated in many cases beyond capacity. That year, as many as half of the 2,400 lagoons leaked, contaminating groundwater. Concentrated Animal Feeding Operations (CAFOs) are large factory farms that house large numbers of animals in buildings, creating efficiencies in feeding and disposing of waste. Meanwhile, they cause extreme hazards for the environment and health of those in surrounding areas and emit enormous amounts of ammonia gas, which returns to the waterways in rain, resulting in explosions of algae growth.

Eastern North Carolina's sandy soils and shallow water table make it especially vulnerable to groundwater pollution, yet the state has the weakest regulations of any major hog-producing state, including Missouri, Iowa, and Virginia, and minimally enforces the laws on the books. Lagoon spills are common, contaminating streams, rivers, and groundwater. Sometimes discharges are deliberate, when, for example, farms run out of places to spread or spray waste.[21]

Today, the majority of the hog industry is located in southeastern North Carolina, adjacent to or near the Tar Heel plant. The top four hog-producing counties in the state are Bladen, Robeson, Sampson, and Duplin Counties.[22] These counties produce more phosphorus-rich manure than the land can absorb.

"Is hog waste treatment up to the same standard as human sewage treatment? Certainly not. And if it's nowhere close, then you've got a mammoth potential problem on your hands," said Lawrence B. Cahoon, professor of biological science at the University of North Carolina at Wilmington, in an interview with the *News & Observer* in 1995, when there were only 2 million

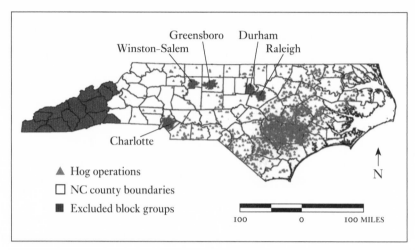

Greensboro   Durham
Winston-Salem            Raleigh

Charlotte

▲ Hog operations
☐ NC county boundaries
■ Excluded block groups

N

100        0        100 MILES

Hog operations sprang up around the Tar Heel plant, increasing the number of hogs in North Carolina from 2 million in 1995 to more than 10 million in 2016. Forty percent are located in Duplin and Sampson Counties, immediately to the northeast of the Tar Heel slaughterhouse. Figure from *Environmental Health Perspectives* 121: A182–A189 (2013), http://dx.doi.org/10.1289/ehp.121-a182.

hogs in the state.[23] A decade later, in 2016, there were more than 10 million, with 4.5 million hogs in Duplin and Sampson Counties alone, producing 4 billion gallons of waste a year, 40 percent of the state's total. Duplin County itself produces twice as much sewage as humans do in the entire New York City metro area. Of course, in New York City, the sewage is treated.[24] "The big integrators could not produce pork chops or bacon cheaper than traditional family farmers unless they got around environmental laws," Kennedy wrote in the preface to *Righteous Porkchop*. "Since the industry produced far more waste than it could profitably dispose of, its entire business plan was based on its capacity to illegally dump industrial scale quantities of raw animal waste and toxic chemicals. . . . With the exception of coal mining's mountaintop removal, no other industry in America's modern history had gotten away with systematic pollution of this magnitude."[25]

North Carolina was especially hard hit by such pollution, Kennedy pointed out. Historically famous for some of the purest rivers and coastal waters in the United States, its rivers and streams became choked with algae fed by nutrient discharges from factories and manure contamination from industry farms. "We regarded Smithfield Foods as the chief villain in North Carolina," he wrote, calling Luter "ruthless" in his determination to maximize short-term profits while passing along long-term costs to communities.[26] In June of 2016, Kennedy's organization, the Waterkeeper Alliance, released an interactive map showing the location of more than 6,500 CAFOs and hog lagoons in North Carolina.[27] It shows more than 4,100 manure pits in North Carolina, with nearly 50 percent of them located in Duplin and Sampson Counties alone. These lagoons—holding billions of gallons of untreated pig urine and feces—have a measurable impact on the watersheds where they are located, not to mention their public health nuisance and annoyance.[28]

In 1997, North Carolina passed a moratorium on new lagoon farms that became permanent in 2007. Nevertheless, about four thousand lagoons are still in use. At least fourteen flooded during Hurricane Matthew in 2016, when state officials flew over regions to identify where lagoon waste was being carried off by floodwaters and could pose a public health hazard. Rising sea levels increase the risk of contamination. Even when lagoons are not spilling into surface water, they still carry *E. coli*, salmonella, and other bacteria that can make people sick or even kill them. Bacteria can contaminate groundwater. Nitrogen and phosphorus kill fish and damage ecosystems. People living near lagoons have an increased chance of developing asthma, diarrhea, eye irritation, and depression.[29] Plus, bottom line, lagoons stink. So it's no surprise that they're put in counties where people can't fight back.[30] Duplin County has 522 hog farms. All but two are owned by whites, though African Americans make up 29 percent of the population. Duplin and Sampson Counties, where 28 to 29 percent of the residents live in poverty, host the largest numbers of CAFOs not only in North Carolina but in the nation.[31]

On May 11, 2017, the North Carolina Senate passed into law House Bill 467, which limits the amount of damages property owners can collect when their homes are contaminated by pollution from hog lagoons. The cap is the fair market value of their property, which is often greatly diminished by being near the large industrial hog-raising centers. The governor vetoed the bill in early May after lawyers from hundreds of plaintiffs—mostly low-income African Americans in twenty-six federal lawsuits against Murphy-Brown LLC, the hog division of Smithfield Foods—submitted evidence that fecal matter from the hog operations had ended up outside their homes and was probably inside their homes and even in their food.[32] The bill was carried by state Republicans, who have benefited significantly from campaign contributions from the commercial hog-farming industry and the North Carolina Pork Council.

---

But a successful slaughterhouse requires more than lax environmental regulations for the hogs. Slaughterhouses use vast amounts of groundwater and discharge millions of gallons of wastewater every day. In the 1990s, Smithfield Foods was already under investigation by state and federal agencies for polluting the Pagan River in Smithfield, where plants were discharging untreated wastewater, and environmentalists called its CEO "Luter the Polluter." The Pagan had been closed to shellfishing since the 1970s, and swimming posed a health risk.[33] Luter was angry about his treatment by both state and federal officials there. "I'll spend $5 million in attorney fees before I'll pay $3 million in fines," he said.[34] Despite continuing a gloves-off fight with both state and federal government, Luter probably knew it was a losing battle. In any case, he was ready to look for fresh hunting grounds.

Perhaps North Carolina taxpayers should have been paying more attention to Luter's problems in Virginia and asked themselves why he wouldn't locate his crown jewel in the state where he was born, raised, and built his business. But Wendell Murphy did not make his reputation looking out for the taxpayers. He had virtually hog-tied state environmental regulators, paving the way for Luter to build his plant virtually unopposed. North Carolina welcomed him, offering a site on the already polluted Cape Fear River on top of a large natural aquifer. The perfect spot. Smithfield did have to get permits to discharge up to 3 million gallons of treated waste into the river every day — into a part of the river already rated as "at or near capacity" for absorbing industrial wastes. But state officials put Smithfield on a fast track, allowing the company to skip a required impact study.[35]

This was simply more of the same in a state where nearly all top government officials were in the same business — hogs. Indeed, the 2005 *News & Observer*'s "Boss Hog" series highlights the number of state and local officials who were in the hog business and directly benefited from the lenient laws in place governing the business of hog raising and slaughtering. The profits were — and still are — high on every level, from the city to the county to the state. According to Smithfield critic Robert F. Kennedy Jr., North Carolina was in Smithfield's pocket. "Using strategically orchestrated public deception, campaign contributions and political clout, the meat moguls captured state and federal environmental agencies which obligingly (and illegally) waived environmental permit requirements, routinely ignored flagrant violations, and transformed themselves into taxpayer funded hand puppets aggressively defending every instance of industry mischief. . . . North Carolina became a company town."[36]

## CHAPTER 4

# The Plant Opens; The Work Is Beastly

**W**orkers in the Tar Heel plant, mostly descendants of slaves emancipated after the Civil War, did not have a great deal of choice about where to work.[1] To say that they had joined the disassembly line willingly would be to pretend there were other options in a county scarred by poverty. It was hard, dangerous labor. But to some people in Bladen County, it meant opportunity.

Sherri Buffkin was 23 with a 17-month-old baby when the plant started hiring in 1992. She jumped at the chance to work in the cavernous box room above the plant floor for $7.40 an hour. Her job was putting together boxes and placing them on conveyor belts heading toward chutes down to the kill and cut floors. "Hell, yeah, I did," she said. "It paid more than any other job around. . . . [Smithfield] was a brand new multi-million company. There were opportunities. From the day I started, I started to figure things out. Nobody knew how to do anything."[2]

The plant opened October 1, 1992, and the first hog was killed on October 5. Some one thousand more followed each day thereafter, in a single disassembly line with about six hundred workers.[3] "I was there when the first hog was killed," Buffkin recalled. "It was a complete and total clusterfuck all day. Then we killed a few more, then a hundred, then five hundred. It was cool. It was fun. It was long hours and great money. Everybody was just figuring out how to do it."[4]

In a nutshell, hogs are trucked in, prodded onto the kill floor, stunned or gassed, stuck in the jugular, then hung upside down and bled out. The tendons in the rear legs are severed, and the pigs are doused in boiling water. The skin is scraped and cut for pork rinds. Blood runs onto the floor as severed hogs' heads float overhead, hooked onto conveyor belts. The bodies move down the lines, the organs are removed, and the empty carcasses are

Sherri Buffkin's
Smithfield Packing
ID tag.

sawed in half, beginning the disassembly that will end in hams, pork chops, and chitlins.[5]

Most sections of a slaughterhouse are called by their purpose: Livestock, the wet kill floor, the dry kill floor, the cut floor, conversion, and shipping and receiving. Most of the plant is color-coded. In Tar Heel, workers handle about two thousand hogs per hour, or one about every three seconds. The wet kill floor has no walls, so temperature varies with the weather, routinely in the 90s in late summer, driven up over 100 by the adrenalin, animal heat, and steam from the scald tubs.

The upside-down railroad track carrying the shackled pigs runs in a zig-zag pattern to give the blood time to drain; then the hog goes through a rinse cabinet and is submerged in the scald tub, about four feet deep and a hundred yards long in the shape of a candy cane. There the hide starts to soften in preparation for dehairing. At times, a hog will come unshackled under water, and one of the workers must retrieve it with a long steel hook and reshackle it to the line. It's difficult, heavy work lifting more than 250 pounds of dead weight.

A big washing machine with paddles takes most of the hair off, then the animal lands on a moving conveyor where workers make sure the feet are pointing in the right direction; if they aren't, the worker has to flip over that dead weight. The workers make two cuts to the back of the feet, exposing the heel string. Workers must thrust a gam—a stainless steel hanger about 18 inches long with a pointed hook—through each side of the hog. This is one of the most difficult jobs in the plant, handled by the largest, strongest men, almost always black.

Next, the hog is hoisted up and continues through a 10-foot-long cabinet with propane flames to burn off remaining hair. The hog then goes through something like a car wash, with spinning brushes to remove the remaining loose hairs. Finally, workers on a scaffold shave hair from between the legs.

On the dry kill floor, the dismantling begins when workers cut out the hair pocket between the hooves on all four feet. Next, a worker with a large pair of hydraulic scissors breaks the neckbone, leaving the head hanging by a large piece of skin. Several workers trim the jowl, separating the head from the carcass. The carcass goes to the bun operator, who cuts a circle around the rectum, dropping the rectal cavity into the carcass without breaking it. This is all being done as the line is running very, very fast. Each worker repeats his or her task over and over and over.

Next, workers use a kind of chain saw to split open the hog's chest without cutting the intestines. Workers lay open the carcass, breaking the animal's hip bone with a hydraulic scissors, which resembles a tree pruner. Two stainless steel hooks hold the carcass's chest open so a worker can pull out the internal organs, from the esophagus to the rectum, in one motion. The next worker in line separates the esophagus, lungs, heart, and liver from the intestines, and off they go for processing. The intestines are fed into a 3- to 4-inch vacuum pipe that leads to the casings and chilling area.

The line splits into three parts, one carrying pans full of hog guts, the second the carcass itself, and the third the head, which is checked by USDA inspectors for tuberculosis and other diseases, pulling heads off the line as needed. Sometimes line speeds are so fast that inspectors can only make quick visual inspections; on the Smithfield lines, heads are moving at about thirty-three per minute. The carcass, guts, and head from each individual hog run parallel so that at any point, if an inspector identifies a problem, workers can match the carcass with the head.

On the carcass line, workers split the backbone with a circular saw, leaving only enough skin to hold the two sides of the carcass together. The kidneys remain inside the carcass within a layer of fat that has to be cut open. "Popping kidneys," as this process is called, is one of the easiest jobs on the kill floor. The kidneys are turned to the outside for easy viewing by USDA inspectors, while interior fat inside of the ribs is pulled out by hand. This is a very tough job requiring great upper body strength. The very last job on the kill floor is scraping out the spinal cord with a small device shaped like a spoon.

Next, the carcass goes into the snap chill portion of the cooler, a room similar to a wind tunnel where the temperature is kept at about −20 degrees Fahrenheit. The hog travels down a long tunnel on a conveyor, does a U-turn, and returns, its outer layer nearly frozen. It's then stored in the cooler bays overnight. The entire process, from jugular stick to cooler bay, has taken between five and ten minutes.

The next day, the disassembly continues on the cut floors, which are kept between 30 and 40 degrees to slow the natural deterioration of meat owing to the growth of micro-organisms like bacteria. The sound of machinery drowns out conversation. Employees wear smocks, gloves, rubber shoes, hairnets, and color-coded helmets to identify department or status: new, probationary employees wear green helmets; regular workers wear yellow; and supervisors wear white ones. On the deboning lines, workers wear safety gear, including metal mesh covers on their forearms and hands, while they hack away meat from the bone. Accidents are frequent.

First, workers cut off the feet and knuckles. From there, the lines multiply, with one for each product: hams, butts, picnic shoulders, loins, ribs, bellies, and fatback. Each line splits off at about a 90-degree angle so that the products can be trimmed according to customer specifications. Some products are packaged, vacuum-packed, and sent to a cooler in shipping. Some go to conversion, a department where workers debone the meat and prepare it for special markets such as packaged marinated ready-to-cook meats.

Meanwhile, the animal parts separated from the carcass are also being prepared for market, though not necessarily as food. Hog intestines are processed in the casings area, a job traditionally done by African American women. The first worker, called a puller, stretches out, straightens, and untangles the intestines, which are rinsed and soaked, then tied in bundles and packed in salt ready to hold sausage or hotdogs.

In the head room, workers cut off the jowls, detach the skin and temples, and trim any head meat from the skull. This is also where they open the skull and remove the brains. Hog brains are a staple in Chinese and Korean stir-fry cooking, used as a thickener. They are also eaten in the South. Former fifteen-term congressman Howard Coble of Greensboro contributed a recipe to *Congress Cooks* for pork brains and scrambled eggs. He notes that canned are okay but fresh are better. He also recommends generous amounts of salt, pepper, and butter, accompanied by a biscuit.[6] The skull is also a source of meat used in smoked pork jowl, or cheeks, a staple of soul food.

Finally, everything that hasn't already been processed for use is rendered (to separate the fat from the bone and protein material) into edible and inedible products. The fat is converted into products like lard or tallow, while the protein is heated, then ground into meat and bone meal, which is often sold for animal feed and fertilizer. As the old joke goes, everything but the squeal is packaged and shipped.

In the box warehouse, workers construct shipping boxes. Packaged product from the cut floor runs on a conveyor belt upstairs to the scale room, where workers pack it in dry ice, seal it, label it, and weigh it. Then it's sent to shipping and receiving. Seventy-five doors open to load products from the plant onto waiting trucks and to receive needed supplies—welcoming opportunity.

---

After only six months working boxes, Sherri Buffkin was making $10 an hour as crew leader and had her sights set on managing all of purchasing, as well as shipping and receiving. Attractive and—like the top bosses—white, Buffkin was on a fast track, passing everyone else to report directly to the white men at the top. That put her on a collision course with seasoned meatpacker—and former union man—Larry Johnson as they both navigated the chaotic early days at the plant and vied for favor with the bosses.

Johnson, a tall, good-looking man with a big smile, was a man's man, ambitious to one day be the plant manager in Tar Heel. He first saw the market shifting while working as a union meatpacker in Milwaukee in the 1970s and '80s for Patrick Cudahy, a Wisconsin subsidiary of Smithfield (and the official bacon of the Green Bay Packers). At least twice he had turned down opportunities to advance to management because his wife was a strong union supporter. "Wages started going down and basically I thought I seen [*sic*] the handwriting on the wall. I sat down with my family and basically said I need to see if I can make some advancements in management."[7]

In 1985, Johnson left behind his fifteen-year membership in the UFCW, where he had been a union leader, helping organize fellow workers as department steward, assistant steward, and occasional chief steward, and had played a key role between local union members and management. When he transitioned into management at Cudahy, he became a non-unionized supervisor, whose first loyalty was to the company, then a superintendent of the cut floor. A union strike, lasting more than two years, was still ongoing when he left for Smithfield Packing, in Smithfield, Virginia. There he was front-line supervisor of fresh-meat grading in the converting room for two years, then a line supervisor for two years. "I heard a rumor that they were considering building a brand-new plant in North Carolina," Johnson said later. "I thought it would be a great opportunity."[8]

He was about 43 when he moved to the Tar Heel area in April of 1992 to plan the layout of the cut floor while the plant was under construction. The biggest challenge was to figure out how to train workers who had never worked in a slaughterhouse. He bought a pretty one-story brick home with a carport on Woodhouse Drive in Elizabethtown—a good, wholesome place for him and his wife to raise the boys.

The opportunities at a new plant also attracted the attention of other high-ranking supervisors within Smithfield, including Jere T. Null. Null had been working for Smithfield at headquarters since fall of 1989 as a financial analyst. The Dallas native had graduated from Old Dominion University in Norfolk, Virginia, in 1986 with a BS in finance and international business. In early summer 1992, before the plant opened, Null accepted a position as technical service director—basically comptroller. He was ambitious and smart.

Director of security William Daniel "Danny" Priest would be another key player. Priest was hired in June of 1993 to help maintain order among the sometimes rough crowd of laborers, a mix of blacks (mainly), Lumbee Indians, whites, and Latinos unused to the rigors of a slaughterhouse line job. Before starting the job at Smithfield, Priest had worked as an undercover narcotics investigator on a federal, state, and local drug task force from 1988 to 1990, then went to the State Bureau of Investigation Academy to be certified as a drug abuse resistance education officer. He subsequently taught Drug Abuse Resistance Education (D.A.R.E.) to middle and high school students in the Bladen County public schools.[9]

Priest, at that time about 38 years old, combined his job at Smithfield with a position as an auxiliary Bladen County deputy sheriff. Because of his power to make arrests, he aroused fear among the workers, Buffkin recalled. "People were scared to death of him. He was their [the company's] thug. He could arrest you and have you fired on the spot."[10] A short white man, about five feet five and a half inches tall and weighing 160 pounds, Priest initially oversaw twelve security guards. They wore white shirts, dark or navy blue trousers, and a small badge. That force grew to twenty-four full-time guards by 2000, when the state certified it as a private police force. Officers could carry guns on duty, make arrests, and interrogate workers at holding cells inside the plant.

Two other hires would play key roles for the union side as the plant expanded, one white, one black. They couldn't have been more different, but they both earned the respect of other workers and become leaders within their areas of the plant. One was Keith Alan Ludlum, hired in September of 1993, fresh from the armed forces after serving in Desert Storm. He'd recently married and returned to his home state looking for work. Broad-chested and muscular, he took one of the most difficult jobs at the plant, driving hogs into the plant. Ludlum worked first shift in Livestock with between twenty and thirty other workers, almost all African American. "I was inexperienced and young," he recalled. "I didn't really understand about unions."[11]

The other was a corn-rowed prankster, Rayshawn Ward, who didn't take shit from anybody but was so good-natured about it that no one could stay angry with him for long. He was puny, short, and scrawny compared to Ludlum's brawny build. But he was as quick on the fast-moving line with his knife as he was with his wit, and that appealed to co-workers and managers alike.

---

As Smithfield Foods assembled its workforce at Smithfield Packing, like most non-unionized corporations, it was leaving nothing to chance. Even before opening its doors, the company hired North Carolina's oldest law firm, Maupin Taylor Ellis & Adams, in Raleigh, North Carolina. Its job was to keep the union out of the Tar Heel slaughterhouse, and the company sent labor attorney William P. Barrett to work on-site as the principal legal advisor to managers on how to keep the union from gaining even a foothold.[12] With his blond hair, blue eyes, and tall, square-shouldered stance and educated articulation, Barrett cut quite a figure among an ill-trained workforce, where even most managers were somewhat rough around the edges. In his typical sass talk, Rayshawn Ward called Barrett a "red-carpet silver-spoon pretty little white boy."[13]

From the company's perspective, hiring a law firm to advise managers on how to "legally" keep unions out was perfectly reasonable. It was simply how it was done, business as usual. But such firms—and their attorneys—are known by unions and businesses alike as "union busters." The business of union-avoidance consulting—which goes by many names, including the Orwellian term "labor law"—started as early as the 1940s, and by the 1990s, companies were paying an estimated $200 million per year to consultants, who today earn some $4 billion. Its techniques are extremely effective.[14]

Typically, union-avoidance consultants use opinion surveys, supervisor training, employee roundtables, employee complaint procedures, job evaluations, and incentive pay to maintain complete control of the workforce.[15] These things may sound inoffensive, but they are designed to frighten and intimidate workers to keep them away from unions. Consultants train supervisors to tell employees about the personal economic costs of strikes, violence, and picket lines and to warn employees that unions will cause them to lose their jobs and their homes.[16] They will sometimes cross the line, illegally advising supervisors on how to identify employees' weak points and target them for what appears to be firing or discipline with cause but is in reality punitive action in response to union support, designed to put other employees on notice that supporting the union is not such a great idea.

Theoretically, Barrett's job was to keep the company on the legal edge of the national labor law, which prohibits firing or penalizing workers who support unionization. But that legal line in the sand was fluid and seemed to constantly move. In the early days, Barrett was the key figure in training supervisors and managers—between fifteen and twenty-five at a time about once a week—on how to conduct themselves during a union-organizing campaign. He was familiar with TIPS training, an acronym used to help managers remember the illegal acts of threaten, interrogate, promise, or spy. In reality, managers also learn to identify employees likely to support unions or to engage in union organizing. To do that, managers often cross the line into illegal activity, as they did in Tar Heel.[17]

Under the National Labor Relations Act, passed during Roosevelt's New Deal, it is illegal for employers to threaten to fire an employee or close a plant if a union is elected. Employers cannot promise rewards to employees who vote against the union. They cannot conduct surveillance of union activities. They cannot interrogate employees about union activities or ask if they have signed a union authorization card. Once the union has gathered enough signed cards, the National Labor Relations Board will require the company to hold an election. Alternatively, the company can directly accept the union without a vote, but this rarely happens. Choosing the election route gives a company more time to intimidate workers and convince them not to vote for the union.

Barrett took the job with Smithfield, ironically enough, right after a four-year stint with the NLRB—the very government agency tasked with making sure employers and employees don't break labor law. Right out of law school in 1988, Barrett worked as a field attorney for four years for the NLRB in St. Louis, Missouri, near his alma mater, the University of Missouri at Columbia.

---

The National Labor Relations Board is the federal agency charged with enforcing the 1935 National Labor Relations Act, which protects the rights of workers to band together to improve wages and working conditions. It also—theoretically—protects their right to choose to have a union represent them in bargaining with their employers. It also—theoretically—prevents companies from engaging in illegal unfair labor practices and provides remedies to workers when companies break labor law by bullying them or firing them if they show an interest in or support a union.

The problem is that all five NLRB board members are presidential political appointees—confirmed by the U.S. Senate—who can independently determine which laws to enforce and when to enforce them, although ultimate power rests with federal appellate courts. As a result of historical shifts along party lines, board decisions range from pro-employee to pro-employer, depending on the whims of politicians. A somewhat stabilizing force has been the tradition of the president appointing three members of his own party and two of the other. Another is the staggered five-year appointments of the five board members.

The NLRB is both an investigative body and a judicial body, but it has no power to enforce its own rulings. While many companies accept those rulings, those who do not—like Smithfield Foods—can appeal the rulings all the way to the U.S. Supreme Court. This process can take many years, leaving workers without remedy for lost jobs, lost pay, or other injuries, until all appeals are exhausted. To give an idea of the sometimes wholly dysfunctional nature of the organization, between 2007 and 2010 the board dropped to only two members—one Democrat and one Republican—as President George W. Bush wrangled with Senate Democrats over appointees, then President Barack Obama with Senate Republicans. The two-member board —Wilma Liebman and Peter Schaumber—issued more than four hundred decisions, mostly upheld on appeal in the U.S. courts of appeal. Then in June of 2010, the Supreme Court invalidated all the rulings. On July 16, 2013, the Senate's filibuster of all President Barack Obama's appointees finally ended.[18] By August there was a full complement of five board members.[19]

While the reasons for it are multifaceted, the NLRB is—in a sense—slowly going out of business. Intake of cases is down 58 percent since 1980, dropping from 44,063 in 1980 to 21,394 in 2014. Published decisions during that time dropped by 85 percent, while the number of NLRB employees (based on full-time equivalents) dropped 30 percent. In 2013, the NLRB reduced its number of offices from thirty-two to twenty-six.[20]

Nevertheless, the NLRA remains in place, laying out a relatively clear path for workers who believe their rights have been violated. First, they, or members of the union they are organizing for, contact the investigative arm of the NLRB, which then can issue—or refuse to issue—a complaint against the employer. The employer has ten days to respond. The charges are then adjudicated during a hearing before an NLRB administrative law judge. The judge's decision can be appealed by either the employee or the employer to the full NLRB board, then on up the chain of the federal court system until appeals are dropped or the Supreme Court makes a determination.

---

Of course, union-busting legal consultant William "Bill" Barrett knew firsthand exactly how the NLRB worked. As a field attorney in government service, he had investigated unfair labor practice charges, taken affidavits from workers and witnesses, conducted NLRB elections for workers to decide whether they wanted to be represented by a union, and represented the NLRB in hearings before administrative law judges to provide evidence pertaining to charges that the union—or the company—had violated national labor law. Usually it was the union bringing charges against the company.

Whatever advice Barrett was giving Smithfield managers behind closed doors, they didn't seem too concerned about the NLRA regulations, openly flouting them, often brazenly breaking federal law. Once the UFCW launched its organizing campaign, head of security Danny Priest made a habit of cruising by union meetings, writing down license numbers, and letting workers who attended know that he knew who they were. It's what's known in the business as illegal intimidation. And Priest and his crew—usually security guards or off-duty sheriff's deputies, sometimes in uniform, at times wearing visible firearms—also enjoyed harassing union workers, especially Gulf War veteran Keith Ludlum, who worked in Livestock, next to the wet kill floor.

Keith Ludlum got to the slaughterhouse well before sunrise every morning, where workers started running hogs at 5:50 a.m. The animals arrived on two kinds of trucks, one with a "belly on the bottom, that's a 200-hog truck," Ludlum explained, describing an open-slatted transport truck with an extra holding compartment close to the ground between the front and rear tires. "Any others usually are 150- to 170-hog trucks." The trucks unloaded into stalls on the A side and B side. What was unique about the Tar Heel plant, according to Ludlum, is that it was "a two-chain kill plant. Most just run one line. Reason we're the largest is we run two."[21] Each stall would hold about fifty hogs. "That way if anything happens, you're not working with 200, just 50." Surrounding the stalls were concrete walls with an alleyway. Special foggers created a mist, cooling down the hogs as they were unloaded.

In the beginning, before People for the Ethical Treatment of Animals (PETA) had released tapes of hogs being mistreated and the "hog-whisperer" changed the way hogs were driven, the atmosphere was a free-for-all, Ludlum said. Hog drivers had canvas straps on wooden handles about 30 inches long. "The sight and sound is more important than touch to a hog. . . . So, we would slit the sides where the canvas flaps would hang free and when they get wet, you can crack it like a whip and make that snapping sound and fifty hogs come shooting down the alleyway away from that sound."

Two electric gates swept the hogs into the killing room. The job could be nasty, with lots of manure. Workers wore coveralls, but pigs got excited coming off the truck and urine and excrement flew everywhere. When it was time to drive the pigs into the killing room, the foggers were turned off. The pigs were squealing. The noise was intense, but everybody had ear plugs.

Livestock workers started the show. "Fifty hogs are like fifty NFL running backs. You're scared shitless. All of us have had close calls." No one wants to get trampled by fifty spooked hogs, each averaging 285 pounds. Hog drivers are susceptible to taking out their frustrations on the hogs. "If you had a struggling hog, blind or deaf, if they broke from the herd and spun around, he's panicked, there's nothing to guide him," Ludlum said. "It's 110 degrees with the heat index in August and humidity and you mess with those ornery things all day long and you ain't making no money and you're mistreated by management, you can transfer that to them hogs real quick. You flip that (hog driver with canvas straps) and catch them across that snout. We also had electric prodders. . . . We used to have a hog shocker that could move a crippled hog. There was a lot of animal abuse in 1993 and 1994 before PETA sued."

The goal is to keep the hogs constantly moving so there's no gap. A gap means money is a-wasting. But keeping all the hogs moving at the same pace can be extremely difficult, and if the ones up front start to panic and screech to a halt, the ones in back keep moving and start piling up. "It's dangerous. If you can't unjam them, you gotta jump the wall to where they're jammed and start them moving again." Sometimes, just to get under management's skin, the drivers would jam the line up on purpose. "Sometimes I'd run fifty hogs into a pen meant for twenty-five and they'd run on top of each other and stack up and by the time they processed two runs of hogs, I'd have some breathing time and could smoke a cigarette. It was crazy."

"You see death daily," Ludlum observed. "It's hard to imagine, but your human psyche changes. You get used to seeing people get hurt and hogs dying and you build up a thick skin. It's amazing what human beings can get used to." He expanded on this theme when interviewed for the film *Food, Inc.* "[Smithfield employers] have the same mentality towards the workers as they do towards the hog. The hog, they don't have to worry about their comfort or whether they are raised humanely because they are temporary. They're going to be killed. . . .They have the same viewpoint to the workers.

They are not worried about the longevity of the worker because to them everything has an end."[22]

In the early years, hogs would enter the kill floor single file, where a worker stunned each electrically and slid it down onto the stick pit table. Here another worker, called a "sticker," would cut into the hog's jugular vein. At the same time, another would put a noose chain on one back leg and hoist the hog up so gravity could help the heart pump blood out of the pierced jugular into a trough that channeled the blood to the plasma room.

As in Livestock, animal abuse was rampant. Injured hogs or hogs not properly stunned could be vicious. The constant danger brought out the worst in workers.[23] One sticker at a Morell plant said that he killed about nine hundred hogs per hour, fifteen per minute, or one every four seconds. "It wouldn't be that hard if they were stunned right. But when most of them are fully conscious, kicking and biting at you. . . . I'd say them motherfuckers who do kickboxing, karate, tae kwon do, they got nothing on me."[24]

One Tar Heel sticker saw lots of problems with the four-pronged stunner originally used at the plant. "If you're killing 16,000 hogs a shift, those guys aren't going to stun all them hogs all the time. Some hogs come out kicking and raising hell."[25] If a hog wasn't stunned correctly, the shackler was supposed to let it drop to the floor, where it could be re-stunned, "but the supervisors don't want the shacklers to do that. . . . If the shackler drops too many hogs, they write that shackler up. A shackler out there don't have no choice *but* to hang hogs alive in order to keep his job."[26]

With live hogs, the Tar Heel sticker said, he sometimes couldn't stick them because they were kicking too much. In some cases, workers hit the hogs on the head with pipes to knock them out, he said. In some of the most troubling stories, hogs were still alive as blood drained from their necks and they hit the scalding tubs, sometimes drowning or being boiled alive. "Because of the line speed, the sticker only gets one chance to make a good stick hole. . . . Bad sticks usually don't have a chance to bleed out. [The hogs] end up drowning in the scalding tank . . . hit the water and start screaming and kicking."[27]

Later, the stunner system at the plant was replaced by gassing. In the gas system, groups of hogs are sealed in a chamber that fills with carbon dioxide. The pigs lose consciousness within about thirty seconds, then are hoisted, where workers slit their throats so they bleed to death. While some say it is still a cruel way to put hogs to death, the method eliminates the struggle between a hog that doesn't want to die and a worker that doesn't want to lose his job.

It wasn't the mistreatment of the hogs that turned Ludlum against management and got him talking to early union organizers. It was the mistreatment of the workers, particularly one co-worker, an African American man in his fifties, who was trying to move a dead hog out of the alleyway that surrounded the pens. The alleyway was three to four feet wide and provided a space to separate out injured or dead animals. The worker was using an electric pallet jack, a motorized lifter with handles. When he got to the corner, where there was a little more room, he tried to turn around but the throttle got away from him, and the jack turned around and caught his leg, crushing it. "When you're turning, you're already in a bind. You can panic. It don't move fast, but it's heavy with all those motors," Ludlum recalled.[28] Smithfield executives said they have no record of the employee or the incident.[29] The next day, when Ludlum got to work, the man was already sitting in the break room, his leg in a full cast, propped up on a bench. He told Ludlum he had to walk from a distant parking space. "Can you imagine with fat and blood on the kill floor, in a full leg cast with crutches? And the management parking lot is 15 feet away? That's the straw that broke the camel's back."

The man had to come to work and sit because the company doesn't have to report an injury if there's no missed work, Ludlum said. If they don't report an injury, they don't have to pay workers' comp. If the worker refuses to come in, he loses his job. "For weeks, I watched this man hobble through the parking lot and across the greasy, wet floors of the kill floor and cut departments to get back and forth to the livestock yard," Ludlum said. When the former soldier tried to arrange parking close to the plant, he was told it was for managers only. "At that moment, a light clicked on for me. Here was an injured man who couldn't get a little help because he was 'just a worker.' Well, I am a worker too. I knew I had to stand up and fight."

By December 1993, Ludlum was an aggressive recruiter who had learned what he could legally do on the union's behalf inside the plant. When co-workers asked if they could get fired for signing the cards, he reassured them that they were protected by national labor laws that make it unlawful to penalize workers for union activity during their breaks. When managers disagreed, he publicly challenged them and told them they were breaking the law. That January, in the break room, Ludlum convinced three women to sign union cards. Supervisors were watching through a square window for about ten minutes. When he came out, they told him to leave company property.[30]

On January 26, Ludlum assured a colleague from Livestock that he could not get fired for filling out the union card because he was protected by national labor laws.[31] A crew leader contradicted him. Ludlum confronted the man.

"Do you know you just violated labor laws?" Ludlum asked. "It's illegal for you to say that."

"You can't do it on company time," the manager responded.

"We're on our time. We're on break," Ludlum said.

"Then you can't do it on company property."

"Yes we can," Ludlum replied.

Supervisors cracked down, coming into the break room to tell workers they would be fired for filling out a card or that having a union would force Smithfield to close the plant. It worked. Ludlum started having trouble getting union cards signed.

Ludlum's boss, Edward Ross Lewis, had been a farmer for thirty years. He wasn't happy about the union organizing. Still, he thought well of Ludlum. "It takes a special type of person to drive pigs," he said, and Ludlum was one of them, aggressive and uncompromising. That attitude would help him become an icon for workers from Livestock all the way to shipping.

## CHAPTER 5

# The First Union Vote; The NLRB Investigates

**M**anagement couldn't overtly fire workers for union organizing, but it was easy enough to drum up other reasons in the chaotic atmosphere of a slaughterhouse, where inexperienced workers came and went as they pleased. Unused to the incredibly difficult and dangerous work, workers often showed up late or not at all. Workers would stay until they got sick of the work, quit for a while, then come back. Turnover was rampant and record-keeping was erratic, incomplete and disorganized. Human Resources was not yet computerized and everything—supervisors' notations of attendance, disciplinary actions, firings—was kept on paper in cabinets and eventually boxes in a warehouse.

Henry Morris, the first vice president of operations, the chief site official of the entire plant, remembered the mammoth undertaking of training an unskilled workforce. The tall, dark-haired man was well liked by workers, polite, considerate, and hands on, always out on the lines every day talking to workers. He compared his role to that of mayor of a small town, but he was strictly a company man.

Under Morris, employment climbed to about 1,500 by 1994. Most of the workers were black. Fewer than 10 percent were Latino, and a small percentage were native Lumbee Indians. Managers and supervisors were white. "We had to hire a thousand employees that had no meat processing or packing experience and train them and establish a stabilized work force," Morris reported. "Absenteeism I think was probably the biggest single problem that we had to deal with in the early days."[1] Indeed, about half the employees had attendance problems. About 9 percent of the workforce were terminated every month.

The reasons for these difficulties are fairly obvious: the rural nature of the area and lack of a large population center, requiring long commutes for some employees, together with the unpleasant nature of the work. A high turnover rate, exceeding 100 percent, is not unusual in the meatpacking industry, and Smithfield was no exception. Between 1992 and 2001, the plant had already employed more than 20,000 people, most of whom had come and gone. Annual turnover was more than 95 percent in 1994 and 1997, and those were good years. The rate was more than 100 percent for 1993, 1995, and 1996. Even the supervisor turnover rate ran as high as 50 percent.[2]

Smithfield was so desperate that it hired prisoners by the hundreds from a prison in Lumberton. In fact, the company became the largest employer of inmates in the state, unnerving workers and residents alike, especially when a prisoner escaped. One, serving two life sentences plus thirty years for murder, kidnapping, and robbery, simply ran away. Another, convicted of killing his mother, escaped and was recaptured within hours. Most were serving time for larceny or nonviolent crimes.[3]

Hiring, training, and retention were such a problem that Morris hired a human relations specialist in June of 1994 to help with the logistics, including writing job descriptions, conducting hundreds of interviews, and managing orientations and training for new employees—with the goal, of course, of reducing turnover. But hiring an HR (Human Resources) person didn't solve the problem. Supervisors tried to keep track of absenteeism and send reports to HR. Human Resources tried using a point system in which workers were warned, given last chances, and fired if they got too many points for being late or missing work. But that failed, because supervisors were reluctant to fire trained workers even if they missed too much work. And it wasn't always supervisors who ignored policy. HR itself was hard-pressed to let people go when bodies were so desperately needed on the lines. Sherri Buffkin—by now in management herself—had one employee she repeatedly tried to fire. Each time Buffkin took her to HR, "once again she received a final warning, a final total warning, and a final, final, final warning."[4] Each time, the employee wound up back on the job.

Still, many workers were fired or quit, turning the HR office into "Grand Central Station," the lobby packed with people waiting for termination checks.[5] Clearance interviews were required but often didn't take place. There were no complete records to quantify why people quit or were fired. If workers were ardent supporters of the union, however, that was another matter. Smithfield managers made sure they knew who the union supporters were, so they could nip the organizing effort in the bud. In such cases they generally made sure to file proper paperwork documenting wrongdoing with HR; sometimes they made up reasons and falsified documents.

The easiest way to identify union supporters was when they passed out handbills in front of the plant, tried to get workers to sign union cards, or attended functions at the union trailer in downtown Tar Heel, a stone's throw from the single crossroads marking the center of what could barely be called a town (since 1990, its population has ranged from about 60 to 150). The United Food and Commercial Workers Union — including professional African American organizers hired from out of state — had set up shop in the trailer about two miles south of the plant, tucked behind the only bank, next to the Texaco station. Across the street was the Goodyear tire store and the Log Cabin Restaurant, with its dark-brown, hand-hewn logs. There, in plain view, the UFCW representatives hosted cookouts and encouraged worker volunteers to get union cards signed.

Upper management was determined that the union not get enough signatures to file a petition with the NLRB asking for an election. One employee they were watching closely was the hog driver Keith Ludlum. "The minute you showed you were a leader you were targeted," he said.[6] Ludlum was fearless and frequently handbilled on Highway 87 right where every employee drove into the parking lot. Because the hog drivers started early and got off before the rest of the first shift, Ludlum could catch first-shift workers as they were leaving and second-shift workers as they were coming in. Handbilling was a surefire way to get on the company's hit list, but Ludlum didn't care. "I was in my prime, six-foot, 180 pounds straight out of the military. And Danny Priest and these little pipsqueaks trying to antagonize us into a fight, it was all I could do to keep from dragging them down the highway."[7]

During the early morning and midafternoon shift changes, it was not unusual for cars to back up a quarter to a half mile as a long two-lane driveway spilled hundreds of cars onto Highway 87. In July of 1993, the company posted a four-by-four sign facing the highway: "No Trespassing. All persons and vehicles entering/departing are Subject to Search. Solicitation & distribution of literature which is not authorized by the Director of Human Resources is prohibited." There, on either side of the property line, union supporters vied with company supporters, both trying to press handbills into the hands of drivers and passengers. Tempers flared.

"You don't need to take that union shit. Roll up your windows, don't take their union material," said Smithfield employee Timm Pridgen, pushing company literature into the cars and snatching union literature out. Pridgen and Priest — the armed head of security — would often lean in or squat down so they were face to face with drivers, an intimidating move. "I know you, yes, I know you. I remember you," Priest would say to employees who took union materials. When one woman took a union flyer, Pridgen told her, "Fuck you anyway!"

Pridgen also loved to taunt union reps, Smithfield workers later reported. "You fat fucking farm boy. Why don't you go back where you fucking came from?" he said to one union rep, calling another "skinny motherfucker" and a third "punk fucking Italian." Union reps would throw insults right back.

On October 21, 1993, a payday, handbilling started about 1:30 p.m. and was in high gear by 3:30 p.m., when the bulk of the traffic hit and cars backed up by a half mile. Workers just wanted to get out and to the few banks in town to deposit their checks. Company employees yelled out to report the plate numbers of employees who wouldn't take company literature. "Did you get that?" one shouted.

The union complained of dirty tricks, including company employees boarding prison buses from the Robeson Correctional Institution and keeping them from taking union handbills, or motioning cars quickly onto the highway. "Come on, keep coming, keep coming," they would yell at the drivers, waving their hands.[8]

Along with handbilling, the company kept track of who attended cookouts at the union trailer on Thursday paydays, which often attracted more than fifty plant employees who were encouraged to fill out union cards. On September 9, 1993, African American laundry worker Lawanna Johnson attended. She had experience as a boxer in backrib packaging, as a breastflap skinner on the shoulder line, and in conversion sanitation, cleaning up meat scraps from the floor. Three or four other women were filling out union cards when a supervisor scolded them. "You should be ashamed of yourselves, over there filling out the cards." Danny Priest also came by that day, and two other supervisors drove by.

Nearly two months later, there was retaliation. On November 1, 1993, at the end of her break, Lawanna Johnson was talking loudly with two men waiting by the laundry counter for gloves, smocks, and caps while co-worker Gregory Spann leaned on the counter inside the laundry about fifteen to twenty feet away. Spann, a heavy-set black man who was first-shift laundry crew leader, liked his job and was proud of the way he had reworked the system to process in new employees and process out those who quit or were fired. Larry Johnson, the ambitious supervisor from Milwaukee who had by now worked his way up to superintendent of the cut floor, was leaning back against the counter about a foot away.

Lawanna didn't see Larry Johnson as she urged her two co-workers, "You should vote for the union. It will make the plant better." When the two men turned around and saw Larry Johnson, they took off. Johnson stuck his finger in Lawanna's face. "If I hear one word coming out of your mouth promoting this union thing, I will fire you on the spot," he said. Lawanna turned to Spann. "Did you hear what he just said to me? They can't tell me what to say when I'm on my own time." Spann was furious. "I heard it," he said, turning to Johnson. "You don't have the right to tell her she can't talk about the union if she's on her lunch break." Larry Johnson exploded. "Shut the hell up and mind your own business." A few days later, Spann was fired. "I was railroaded out of there," Spann said.[9] Three days later, Lawanna Johnson had to rush her husband to the hospital with respiratory problems. She called her boss but was promptly fired for missing work.

---

Spann and Johnson weren't alone. The pattern was becoming clear to union organizers. Workers they recruited and groomed to be internal union leaders were being targeted and fired. Next to go was George Simpson, hired in November of 1992 as a neckbone puller and later moved to become a picnic trimmer. He was a hard-core recruiter, leafleting hallways, the cafeteria, locker room, break room, and parking lot. He talked to employees in the bathrooms, the parking lot, the cafeteria. He wore union T-shirts and union buttons to work. During his hiring orientation, a manager had asked if anyone had prior union experience. "I was a union shop steward in New York City," Simpson told her. Director of Human Resources Sherman Gilliard told him, "We don't want a union in our plant. We don't need a union taking money out of our pockets." Gilliard was the only high-ranking black manager at the plant.

About three months later, Simpson attended a mandatory anti-union speech given by plant officials for about two or three hundred employees. Unions call them "captive audience meetings" because they can be coercive and intimidating. They are legal as long as the information is factual and no threats are made, though unions aren't given equal time and workers are significantly more likely to vote against a union after attending captive audience meetings.[10] Smithfield held them during extended lunches or morning or afternoon breaks in the cafeteria.

Plant manager Henry Morris would give a rehearsed speech from a typewritten statement, reminding workers how competitive and highly regulated meatpacking was. "You would be smart to look around you and see who these people are that talk the union up. Ask yourself, are you willing to risk your job and the future of this operation on them? One thing is certain, if the going gets rough, they ain't going to be around to help you pay your bills or find you a job."[11] Morris recited statistics about the number of pork plants that had closed after unions won elections. "The union has won elections in over 500 pork processing plants across America," Morris said. "How many of those plants where they won elections are open today? The answer is . . . there are only 46 plants open."[12] He warned them that sometimes wages and benefits went down. "The promise that you have everything to gain and nothing to lose is another union lie," Morris said, promising them a raise that September.[13] Simpson took issue with Morris's speech and told a manager that in his experience employees weren't able to bargain effectively without a union. Both of Simpson's direct supervisors were there. Simpson was clearly a marked man.

Simpson had an excellent reputation working on a line pulling neckbones for $8.25 an hour, the second-highest-paying job. He had been recognized for achieving 120 percent production time, that is, handling 20 percent more product than normal. After he spoke out, he was assigned to the picnic trimmer line, which paid $7.65. There he used a straight knife and wore a steel chain cutting glove, a safety apron, and a smock.

On May 13, 1993, Simpson left the line to use the restroom without waiting for other workers to return. When a supervisor tried to stop him, Simpson cursed him out and went anyway. He accepted a one-day suspension.[14] On May 28, he was marked absent, possibly for visiting his mother, who had had a series of strokes. This time he refused to sign a disciplinary document. In June, he was warned about being careless and not wearing safety equipment when he cut his finger. For safety reasons, because he was left-handed, his supervisors assigned him to the dreaded last position on the line so that his left elbow and his knife would not interfere with other workers. The real reason, he later said, was to punish him and set him up. "As the meat came down the line, each man would pull a piece, and the last man's job was to see that no meat went uncut through the machine," he explained.[15] Workers were supposed to rotate through the last-man position because it was the hardest job. "But my supervisor made it clear that no one was supposed to change positions with me in that last spot." That July, Simpson was suspended for absenteeism, apparently for visiting his mother. In December, he was again marked absent.

Simpson continued to stand up for worker rights. On the morning of January 10, 1994, he complained to his supervisor that the line was running too fast and workers couldn't keep up. That day he was falsely accused of refusing to stay late and trim product. "There were no hogs hanging on the chain," he protested. "The last hog comes off the line, the line is finished."[16] That same day, his team was issued a warning for low production. Simpson refused to sign, because their performance had been hurt by the fast line speed. One month later, on February 10, 1994, he was called to HR and told he was being laid off as part of a plant-wide cutback. It wasn't personal, he was told. Yet he saw people wearing green hats, indicating they were still within their ninety-day probation. "Probably ninety percent of my line was green at the time I was terminated."[17] The union filed a complaint about his firing, one of many.

---

The United Food and Commercial Workers (UFCW) union filed its first complaints with the NLRB against Smithfield Packing on June 25, 1993, and NLRB field agents launched their investigation. Officials at the plant were notified that same day about alleged violations contained in section 8 (a)(1) of the law, colloquially known as "Eight-A-Ones." These violations involve threats, like threat of plant closure, or illegally questioning employees about their sentiments toward the union. NLRB investigators have forty-five days from the time a complaint is received to investigate. From the NLRB Winston-Salem regional office, investigators drove two and a half hours into Bladen County. They interviewed workers and got signed affidavits. They interviewed supervisors and took position statements from Smithfield Packing in response to the union allegations.

Four investigators worked the case, most hailing from rural North Carolina. They were tenacious and unyielding, one attorney later recalled, leaving "no stone unturned."[18] One was African American; three were white. All were rooted in southern culture; several loved hunting and fishing. They were big Carolina fans: back at the office they were called the "Carolina Crowd," always ready to lay a $2 or $5 bet on a University of North Carolina basketball game. These investigators sincerely appreciated working people. They went into trailer parks and neighborhoods where residents were suspicious, wary, and sometimes armed. They didn't carry weapons, instead relying on their wit and personality. Soon they would be investigating more serious charges, the alleged illegal firing of workers who were actively engaged in organizing for the union.[19]

Keith Ludlum was on the radar of company officials. "Within four months of going to work there, I had 98 percent of Livestock signed up. Even the butt kissers for management were signed up," he recalled.[20] Workers were tight, Ludlum said, especially back when the plant was only running one shift. "You felt like you were the first guys doing this in this area of a new plant. And things shouldn't be this way for a new site. People wanted to change things, even management's favorites who were bucking to become supervisors. Even they knew things weren't right, so they would sign the [union] card."

Supervisors either weren't paying attention to anti-union attorney Barrett's training, or they were being urged by upper management to cross the legal lines. When Ludlum signed up Livestock workers in the locker room, supervisors would frequently burst in and tell them not to sign, or tell them that if they did sign they could get fired. Ludlum wasn't intimidated by this illegal intervention: "I was a U.S. veteran. Not much scared me after being in a war zone. I would tell them: This is my federal right under the law. I wasn't afraid to aggressively challenge a supervisor and say: You can't do that. That's illegal. They'd say—Oh sorry. I didn't go to the class yet."

Ludlum had read the labor law and truly believed he was protected. He spread the good news, talking to people he trusted, grabbing them after work and telling them how the union would protect them. When co-workers warned him he was being watched, Ludlum quoted national labor laws that protected his right to talk about joining a union. He wasn't afraid, he said.

Like many other plant workers at the time—including the hundreds of employed prisoners—Ludlum had legal issues requiring appearances in court. Before they realized Ludlum was a union man, Ludlum said they looked the other way and told him openly it was no problem. Once he began teaching co-workers in Livestock about their rights under federal labor law, getting off work to go to court suddenly became a problem. "I got back from court and they called my supervisor and he was waiting for me, turned me right around and said they need to see you in HR."

Ludlum sat in the foyer of the plant outside HR. "If you were sitting in that chair everybody knew you were in trouble, like the principal's office, just sitting there feeling stupid." He was told to hand over his badges and leave. Two days later, he was called back in and officially fired for attendance violations. Ludlum argued vehemently against the charges. He had not been late. He had not missed days. He had not been issued his warnings. Even the dates he'd actually been in court were not the dates on the paperwork. But it did no good.

When he walked out of HR, two sheriff's deputies were waiting for him. "Can I help you?" Ludlum asked. They responded by lining up—one on each side—and walking him out of the foyer all the way to his car. "The sheriff's uniform. That's sending a message," Ludlum said. He went home to his wife, pregnant with their first child. He had been blackballed. He couldn't get a job; he lost his car; he couldn't pay his bills, buy groceries, or even get things for the baby.[21]

Ludlum was angry and fired up. Along with dozens of paid union reps and volunteer organizers, he visited workers in their homes, explaining union benefits. By that summer, enough union cards had been signed that the UFCW was able to petition the NLRB, and a union vote was scheduled for August 25, 1994.

---

Meanwhile, complaints of illegal actions by Smithfield were piling up. By June 14, 1994, the NLRB had notified the company and ordered the complaints consolidated into a single case to bring before an NLRB administrative law judge. They included the firings of six people between April 1993 and February 1994, among them Lawanna Johnson, Keith Ludlum, and George

Simpson, and alleged the firings were motivated by the workers' support for the union.

Then, on August 10, 1994, plant manager Morris told about two hundred employees during a mandatory meeting in the cafeteria that if workers —then mainly African American—supported the union, he would bring in Hispanics, arriving in unprecedented numbers in North Carolina as part of the mass migration caused by the implementation of the North American Free Trade Agreement (NAFTA) the previous January.[22] It was a new play in the old game of racial divide-and-conquer. Every time it fired workers who supported the union with impunity, the company was establishing a terrifying example for workers desperate for their jobs. "It's the hangman philosophy. In Virginia, you hang one, the rest will fall in place," Ludlum recalled. "Inside the plant, it was the same." Ludlum confirmed his reference was to lynchings and slavery, a comparison he said was never lost on the black workers who made up the majority of workers in the early years.

The union vote that August would be monitored by the NLRB. About 1,300 out of the plant's 1,600 workers—nonsupervisors hired before July 24 —would be eligible to vote. Workers hired just before an election are not permitted to vote, under the theory that management can fire union supporters and quickly hire anti-union workers to stack the deck.

Still, nothing was being taken for granted. Two workers in the boxing room were being ordered to take time off from their regular jobs and lobby workers against the union. As Sherri Buffkin recalled, "One left the plant, going around to all the businesses convincing them to go against the union. She and I were good friends. The other one only went in the plant." Buffkin said she advised everyone she knew to steer clear of the union.[23] She was keeping her eye on opportunity.

The plant was slaughtering between 14,000 and 16,000 hogs per day, and the company expected to hire 650 more workers by year's end. A $45 million expansion was underway to increase production to 24,000 hogs a day in response to new contracts from Europe and Japan and too much was at stake, Buffkin thought, for the company to give an inch to the union. "An economically devastating strike" would result if the union came in, plant manager Henry Morris told the *Fayetteville Observer*, comparing the Tar Heel plant to Smithfield's Cudahy plant in Milwaukee, where a two-year strike forced the company into temporary bankruptcy, laying off workers.[24]

The Cudahy strike, and the infamous Hormel strike that left thousands of workers devastated, were frequently cited by managers at the Tar Heel plant to show workers how badly things could go if unions were involved. Smithfield had purchased the plant in Cudahy, Wisconsin, in 1984 for $29 million. Management at Cudahy—relying on labor lawyer advice and reporting to Smithfield Foods—was demanding wage cuts and the right to eliminate insurance for retirees and increase part-time workers to 20 percent. When the company asked for a wage concession from $9 to $6.25, the third from a high of $13 in 1982, the plant's 850 employees went on strike on January 4, 1987. By December, the plant announced it was laying off seven hundred replacement workers.[25] The strike ended on April 30, 1989, after workers failed in a last-minute bid to buy the company during bankruptcy.[26] Workers had struggled to survive for two years on $40-a-week strike pay and donated food. The Local P-40 was dissolved after the strike.

Far more humiliating for the UFCW was the infamous P-9 strike at the Hormel plant in Austin, Minnesota. It provided the immediate backdrop for the labor fight at the Tar Heel plant. In the 1950s, the flagship Hormel plant had produced the wealthiest packinghouse workers in the country, fiercely proud of their skills and hard work.[27] By as early as 1963, unionized labor began granting wage concessions, and in 1978, to induce Hormel to open a new plant, the union agreed not to strike for three years after the new plant opened. The new Austin plant opened in 1982 with fewer than half as many workers as the old plant. Old safety features were gone. Worker injuries went up 120 percent.[28]

In August 1985, when the UFCW international union agreed to lower wages at all Hormel plants, local P-9 officials refused to cooperate, and without approval from the UFCW, 1,500 workers walked out, closing the plant down for five months. "If concessions are going to stop, they are going to have to stop at the most profitable company at the newest plant," one local union leader argued.

By January 13, 1986, Hormel had reopened the plant and hired new workers at lower wages. Riots broke out over strikebreakers crossing picket lines. On January 21, 1986, the governor called in the National Guard. The international UFCW ordered the local to call off the strike. When the local refused, the UFCW placed it in receivership and took over. The local felt betrayed and accused the UFCW of paving the way for union dissolution.

To be fair, unions across the country were being forced into wage concessions by industry leadership, unleashed by President Reagan's firing of federal air traffic controllers and a new era of pro-business political power. The UFCW was under tremendous pressure to offer concessions and avoid massive job losses. That pushed the UFCW packinghouse division to maintain an industry-wide negotiating strategy to convince workers to accept lower wages, but the strategy was failing. The philosophy was to "look for long-term, gradual, and incremental gains; postpone the big struggle until times are more favorable."

Meatpacking companies continued their broad attack on unionized workforces, pulling out all the tools they had to scrap union contracts and cut wages and benefits, including plant closings, bankruptcy ploys, divestitures and acquisitions, marketing agreements, and other devices.[29] Spitfire unionizers at the locals believed that the new UFCW simply could not stomach the old-style militant union fight. As a result, they blamed the UFCW when meatpacking companies became even more brazen. Workers across the industry were demoralized by the humiliating defeat of the Austin Hormel P-9ers, and the gutting of slaughterhouse worker rights continued.

The migration of factories into right-to-work states decimated long-established communities who had depended on the unions through the middle part of the century to create good-paying jobs that assured their children would be educated and surpass them in achieving the American dream. Displaced workers were angry not just because they worked in unspeakable conditions, producing hundreds of thousands of pounds of pork products at an intense pace, and not just because wage cuts were an affront to a middle-income lifestyle. Workers believed the industry was undercutting values that they had internalized during childhood in the 1950s and 1960s, part of their personal and class history. The sense of betrayal ran immeasurably deep. An entire unwritten social compact was being dismantled, a compact that had promised that a lifetime of hard work would be repaid with job security. Workers believed they were investing in their nation, their community, and their children. They were not expecting the sleight of hand that companies began to use to line their own pockets directly from the wallets of their workers.

"The decisive factor behind the revolution in the meat industry was the drive by management to decrease the share of the industry's wealth that went to production workers."[30] As they succeeded, the will of the workers seemed to have been broken. But the Hormel fight had made it clear that when workers unite with a common cause and believe that what they are doing is right and the company is in the wrong, a powerful movement can erupt, sometimes without warning.

---

Most workers at Smithfield Packing in Tar Heel did not think about that big picture. They were simply afraid of losing their jobs. But Smithfield managers and UFCW organizers were fully aware of the humiliating union defeats in Wisconsin and Minnesota. Both knew exactly how high the stakes were. The sheer size of the Tar Heel workforce made the plant alluring for a union desperate to show its meatpackers it could still work for them. A victory here would be a critical one. Both sides were willing to pour millions of dollars into the organizing battle taking shape.

Ultimately, the fight was destined to play out along historical southern power and racial lines. The company had support from government officials and business leaders, all white. The union pulled in local, state, regional, and national black figures who understood the historic exploitation of black workers in North Carolina. Civil rights activist and prominent North Carolina clergyman Nelson Johnson summed it up. "A lot of complaints grew out of the working conditions, the intensity of the work, the lack of adequate breaks," he said. "But the supervisors were all white, and acts of blatant racism by supervising staff were a real factor."[31] Those in the trenches knew it was true. "It's a daily practice, just like profanity. It's a hog slaughtering plant," Buffkin said.[32] Another worker agreed. "The word nigger is used like pouring water out of your faucet every day."[33]

Human rights and civil rights became the umbrella under which workers could be heard regarding the high injury rate made worse within the largely untrained and unskilled workforce. Pro-union workers would say, "They're not killing hogs, they're killing people." And while it wasn't literally true, it often felt that way. Temperatures in the plant were extreme. The speed of the line was intolerable. The best-paying job in town couldn't pay enough to make up for the conditions. Workers were fed up with accidental cuts and infections from wielding razor-sharp knives, repeating the same cutting motion again and again every few seconds. They were fed up with the disgusting contact with animals, dead and alive, that would leave them covered in feces, blood, and mucus.

Stories were beginning to spread that reporting injuries or going to the company clinic was the fastest way to get fired and that managers were being pressured not to report injuries to avoid government scrutiny and workers' compensation payments.[34] Line speed was a constant issue. The time from when the hog was stuck to when the animal was flash-frozen was between five and ten minutes. The rest of the dismemberment into specific cuts of meat took only about five or six minutes more. Another problem was equal pay for unequal work. One woman trimmed snouts off thousands of hogs' heads every day for $6.90 an hour, while a co-worker placed the trimmed meats in boxes — far less taxing — for the same amount of pay. Another was paid the same to observe and make sure everything was running smoothly.[35]

Often the divisions in labor ran along color lines, workers complained, with blacks doing the nasty, heavy-lifting labor and whites doing the easier, cleaner work. Historically, black women had to take jobs in casings, where men refused to work, while white women worked in sliced bacon, where it was cleaner. Black men worked on the kill floors in shackling and scalding tubs and in rendering, rarely allowed to take the knife jobs.[36] Similar patterns emerged in Tar Heel, where whites tended to work in maintenance or in the warehouses; African American men drove the livestock and did the slaughtering, while African American women did the dirty work forcing ground meat into intestines and working with chitlins. Latinos ended up working on the cut floor, slicing larger portions of meat into smaller pieces.[37]

Leading up to the August 1994 union vote, white business leaders opposed the union. The plant was the region's biggest employer and nearly single-handedly had reduced unemployment from more than 12 percent to 7.6 percent in the two years since it had opened. Secondary support businesses had sprung up. Local farmers were raising hogs under contract, resulting in the growth of feed and other businesses. As the county's second-largest taxpayer, the plant was expected to pay more than half a million into local coffers that year.

That July, the Elizabethtown–White Lake Area Chamber of Commerce passed a resolution that "union organizing in our area will not foster economic development growth nor improve employee and employer working relationships."[38] Two billboards on Highway 87 said, "NO dues, NO strikes,

CAROLINA FOOD PROCESSOR EMPLOYEES
NO Dues, NO Strikes,
NO Union-Caused Trouble
Vote NO UNION on August 25
Paid for by Citizens For A Better Bladen County

Sign urging workers to vote against the union, placed in front of the Tar Heel Slaughterhouse leading up to the 1994 union vote. *Fayetteville Observer*, August 16, 1994.

NO Union-Caused Trouble, Vote NO UNION on August 25." A small line at the bottom of the billboards said they were paid for by "Citizens for a Better Bladen County."[39]

Smithfield handed out T-shirts. The front showed two hogs dressed in overalls, caps, and heavy work boots. One had its forefoot hooked in the bib of its overalls and was saying, "What do you think of this union mess?" The other hog replied, "Well, I don't want to get BUTCHERED by the union."[40]

The Saturday before the union vote, Reverend Jesse Jackson spoke before about two hundred workers at the Bladen County Community College. Historically, Jackson was the third African American candidate to make a major-party bid for president. He ran in both 1984 and 1988, urging workers to vote for the union "with the intensity of a tent revival preacher." Jackson was deeply committed to union causes, frequently walking picket lines, and was attuned to the plight of meatpackers. He had prayed with strikers in the P-9 strike against Hormel in Austin, urging them to keep their eye on the prize: the return of their jobs with dignity. Jackson waved his hand and shouted, "I am somebody!" while the crowd chanted back.[41]

Meanwhile, Smithfield managers were watching—and targeting. Worker and union activist Fred McDonald, who had gotten more than a hundred union cards signed, spoke from the podium for about five minutes at the

Reverend Jesse Jackson at a pro-union rally in Dublin, North Carolina, on August 20, 1994. Photo by Marcus Castro, *Fayetteville Observer*, August 20, 1994.

rally and appeared on TV. Fellow worker Chris Council sat on stage with Jackson wearing his union T-shirt. Council had started on the line opening hogs in October of 1992. He liked to clown around, daring other workers to compete, to see if they could open every single hog coming down the line instead of taking turns. On this day, though, Council was serious and in awe as he watched Jackson up close.[42] "We are together, black and white," Jackson said. "Stop the violence, save the worker, save the family." Most of the audience were black, but Jackson told them not to make the vote an issue of race. "This is not about black and white. It is about wrong and right," he said. "We are all God's children. Don't let them deceive you. Don't let them play neighbor against neighbor."[43]

Jackson's arguments were in vain. On Thursday, August 25, the union lost the vote, 704 to 587. Union officials said there was a pattern of illegal voting by ineligible workers and immediately predicted that the election would be overturned and a new vote ordered by the NLRB. "We aren't going away," one union rep said. "We're going to be in Tar Heel for as long as it takes to organize this plant."[44]

Outside the union trailer in Tar Heel, about one hundred workers waited for news of the election's outcome. Keith Ludlum was stunned. Some cried. "I thought we had it, I really did," said one worker who cut shoulder blades.[45] Pickup trucks and cars went by, drivers honking horns to taunt union supporters. Some plant supervisors turned around to pass by and honk again. Company officials went to the plant to personally thank workers on the line. "When it got down to their choice, it was clear that a majority chose not to believe the rhetoric and lies the union was saying," plant manager Henry L. Morris said. "I'm ready to put it behind us. We're going to do everything in our power to run this operation smoothly and cooperatively."[46]

Challenging Morris's claim were more than one hundred complaints filed by workers with the NLRB over alleged labor-law violations. Company executives called the charges "sour grapes." This "was probably one of the most fair elections that has ever been held," Morris said. "We made a valid attempt to tell each employee the truth about the risk of joining a union."[47]

Business leaders said that the vote was the best for Bladen County. A plant manager for a local yarn manufacturer, a furniture manufacturer, and the chairman of the Bladen County Economic Development Board all said that they were pleased with the outcome. "That's a world-class company that has come in here and given two thousand jobs to people throughout the region. A new business will make some mistakes, but I'm sure they will do what's best for the employees."[48]

## CHAPTER 6

# Environment and Immigration;
# North Carolina Is Forever Altered

**B**usiness was booming, but local government officials were beginning to realize that the enormous tax and income benefits Smithfield brought were being delivered with a huge price tag. Even before the plant opened, Smithfield Packing was paying more than $100,000 in taxes, and in 1991 it contributed an estimated $11.5 million to the county economy.[1] Bladen County's tax base grew 13 percent in fiscal year 1993/94, to about $1.1 billion. Industry provided nearly 27 percent of the property tax base; the plant alone paid about $752,000 in taxes that year. Two other hog-related businesses were in the top ten, including Murphy Farms, which paid $107,000, and Carolina Cold Storage, which paid $87,000.[2] In 1995, the additional revenue left the county with a surplus of $3.9 million, up $606,000 from the previous year. It was dizzying.

A rising global demand for pork drove continual expansions and improvements. Production expanded to accommodate customized orders for new Japanese consumers, which accounted for 25 percent of exports by 1996. But other countries had preferences as well. "Holland is the top destination for pig tongues. Koreans like pork bellies. Chinese demand for pork stomachs drives the price from 18 cents a pound to as much as 60 cents a pound. Feijoada, the Brazilian national dish, must have a pig tail to be true to its origins as a slave dish cooked from plantation owners' castoffs."[3]

In 1993, the plant launched a new ham boning line, a new loin boning line, and a new conversion room. A blast-chill wind tunnel, where hog carcasses got a ninety-minute frosting in a wind-chill factor of −65 degrees, was added to the kill floor in late 1994.[4] A large distribution center was completed in 1995, and employment spiked to three thousand workers. Unemployment in the region fell from 11 to just over 6 percent.

By August of 1996, Smithfield Foods projected that its Tar Heel plant would account for nearly one-third of the company's $3.2 billion sales and produce about half its exports. The plant's export processing room grew to 200,000 square feet to handle customized overseas markets. "The Japanese have very definite tastes," said Jere Null, who was promoted to general manager when Morris left in 1995. "You can't just take American pork and sell it in Japan."[5] Not everyone was pleased. "The plant has definitely been an economic plus for the county, but it is also taxing local services and the environment," said Delilah Banks, chairman of the Bladen Board of Commissioners. "It is really hard to say if one outweighs the other right now."[6]

Between July and November of 1995, ninety-five new hog operations with two hundred or more hogs had opened, in addition to the thousands already in place. All were spraying waste from their lagoons into open fields that drained into nearby streams. In Bladen County alone, the hog population had grown nearly 700 percent since Smithfield Packing opened. In 1995, a lagoon rupture and spill at an Onslow County farm—the worst hog-waste spill in the state's history—sent 25 million gallons of waste into tributaries of the New River, killing thousands of fish.[7]

The controversy came to a head when the company applied for a permit to increase the wastewater it was already discharging into the Cape Fear River from 3 million gallons up to 4.5 million per day. But the plant was also having a problem getting enough water to begin with. Groundwater levels in the area had dropped more than ninety feet to satisfy its thirst for 3 million gallons per day since the plant opened in 1992, threatening wells and the Upper Cape Fear Aquifer, which provided water through seven wells. The wastewater permit was essential to Smithfield's proposed $30 million expansion, which would bring the plant to its full slaughtering capacity, from 24,000 to 32,000 hogs a day.

When officials delayed approval on the permit in December 1995, calling for studies on the environmental impact, an impatient Smithfield sued the state, alleging it was holding the permit "hostage" to prevent new farms from opening. The suit argued there was no proof that the number of hog farms would increase just because the plant expanded. In fact, it said, the state had it backwards: it was the growth of hog farming in North Carolina that encouraged Smithfield Foods to expand the Tar Heel plant. County officials were beginning to express their concerns publicly. "It's disappointing to lose an expansion, [but] if the expansion is going to put Bladen's environment at risk, then we might need to make do with what we have," said Bladen County Commissioner Delilah Banks. "We need the revenue. We need the jobs. But once you do irreparable damage, it might take a lifetime to correct."[8]

As Smithfield's suit went forward, the state continued their impact study. By May of 1996, the company had provided the state's Department of Environment and Natural Resources with hundreds of pages of impact information. The plant had been cited nineteen times for wastewater discharges into the Cape Fear River between 1994 and 1997, exceeding limits on fecal bacteria, ammonia, and pollutants that reduce oxygen levels in streams and rivers. Smithfield said that most were for exceeding limits by "small amounts."[9] In December of 1998, the Southern Environmental Law Center filed a complaint on behalf of eleven environmental watchdog groups citing Smithfield's numerous violations of its wastewater permit.[10]

The state eventually issued a wastewater permit in 1999 that capped the number of slaughtered hogs at 32,000 a day and capped the amount of wastewater discharge at 3 million gallons a day. Smithfield was permitted to exceed the cap if they bought from hog producers using innovative waste-disposal systems, but were prohibited from buying hogs from farmers who had been fined for illegally discharging waste.[11]

---

The company's environmental damages were not limited to North Carolina. In August of 1997, a federal judge in Norfolk, Virginia, fined Smithfield Foods $12.6 million for polluting the Pagan River, the largest penalty ever levied for a violation of the federal Clean Water Act. Smithfield had illegally discharged wastewater—containing numerous pollutants, including elevated levels of nitrogen and phosphorus—from Smithfield Packing and Gwaltney of Smithfield, Ltd., between 1991 and 1997. Both were located on the Pagan River, a tributary of the James River in Isle of Wight County. Smithfield eventually was forced to connect the wastewater to a government sanitation system.

In 1988, when state regulators originally mandated that Smithfield upgrade its wastewater facilities, Smithfield filed suit claiming it was "technologically infeasible." "Because these new phosphorus limitations were not required in other states, Smithfield also began to talk publicly about moving its operations out of Virginia rather than complying. . . . When it became apparent to the EPA that Virginia did not intend to initiate legal action against Smithfield . . . the EPA filed its own action."[12]

By 1995, state and federal agencies were investigating the chief operator of Smithfield's wastewater treatment plant for allegedly destroying and falsifying records submitted to environmental agencies. (The 45-year-old man later confessed and served thirty months in prison, saying superiors told him to do it, but he later refused to testify against Smithfield when he was not granted immunity.) Joe Luter insisted the man was a "rogue employee" who acted alone.[13]

As a result of their investigation, state officials concluded that Smithfield violated water pollution permits more than 22,500 times between 1987 and 1997, with most violations involving the missing or falsified records. Others included dumping excess fecal coliform bacteria, chlorine, nitrogen, and other pollutants into the Pagan River. These legal developments naturally alarmed environmentalists in Bladen County. A coalition of more than thirty environmental groups began citing the Virginia case in urging the state to investigate the Bladen County plant.

---

Meanwhile, another global shift was taxing local government services and significantly altering the communities around the plant. Extraordinary numbers of Latinos began arriving in North Carolina from Mexico and Central America, many undocumented, driven by the shockwaves of NAFTA. This mass immigration would radically shift the playing field inside the plant, with an enormous impact on hiring practices. The union too would be scrambling to adjust its strategy to encompass the changing face of the workforce. The North American Free Trade Agreement, a treaty between the United States, Canada, and Mexico that went into effect on January 1, 1994, eliminated tariffs on most exports and imports. This resulted in decades of economic failure in Mexico, hurting small farmers the worst. The flood of taxpayer-subsidized U.S.-produced corn into Mexico undercut domestic markets, forcing about 2 million Mexican farmers off their land when they could not compete in either growing or marketing their goods in competition with agri-industrial U.S. companies. Between 1991 and 2007, Mexico lost 1.9 million jobs in agriculture and 4.9 million family farmers were displaced.[14]

Those who had lost their jobs surged across the border, and between 1994 and 2000 the number of emigrants increased by 79 percent. In 1990, there were 4.5 million Mexican natives living in the United States; by 2000 that would jump to 9.4 million, and by 2009 to 12.6 million.[15]

This unprecedented immigration—initially into southwestern states— began to trickle, then flood, into the South and Southeast of the United States, where the economy was robust. Initially, most were young, single (median age 27) men (63 percent) from Mexico, with little education, who were not fluent in English (57 percent). Most were undocumented. Between 1990 and 2000, six southern states experienced the fastest growth: North Carolina, Arkansas, Tennessee, South Carolina, Georgia, and Alabama. By far the highest rate of increase in the nation was in North Carolina. According to the U.S. census, the Latino population in North Carolina grew 394 percent, from 76,726 in 1990 to 378,963 in 2000. In some counties, including those around Tar Heel, growth was much higher.[16] In 1990, Robeson County, near the plant, had a population of about 105,079, with 704 Hispanics, the census showed.[17] By 2000, just ten years later, the number of Hispanics jumped to 5,994. By 2010, the Hispanic population nearly doubled, to 11,028 of 134,068 total population.

This was a boon to employers, who quickly hired thousands of workers for a median annual income of $16,000, about 60 percent of the earnings of white workers, and sometimes as low as 47 percent of that of white workers. While overall poverty in the six states dropped by 7 percent, the poverty rate among Latinos jumped from 19.7 percent to 25.5 percent between 1990 and 2000, with 41 percent of those under 17 living in poverty.[18]

By 2010, the Hispanic population in North Carolina had more than doubled to 890,000, or 9 percent of the state's population. The median age was 24. Median earnings were about $19,000.[19] About half were undocumented.[20] Many came looking for jobs in the rural meatpacking, agriculture, poultry-production, pork-processing, and textile plants that had scattered across nonmetropolitan North Carolina during the southern manufacturing boom. The impact on meatpacking was nationwide. Between 1980 and 2000, the percentage of Hispanic meat-processing workers rose from less than 10 percent to more than 30 percent.[21]

The trend was mirrored in the Tar Heel plant. During the 1994 vote, only about 200 of the 1,600 workers—or 12 percent—were Hispanic, with about 60 percent African American and about 25 percent white. Within three years, about 940 of the now 4,000 workers—almost a quarter—were Hispanic.[22] In counties around the plant, trailer parks sprang up to house the new arrivals, typically featuring chain-link fences, toys strewn along bumpy dirt roads, outdoor grills, and satellite dishes. Groceries and small cinder-block restaurants opened with signs handwritten in Spanish. As more Latinos moved to the region, the union brought in Latino and Latina organizers.[23] At union rallies, stickers, T-shirts, and posters were printed in Spanish. Speeches were given in or translated into Spanish. Likewise, the company began using interpreters and bilingual printed materials. But many of the Latino plant workers could not read or write in Spanish, much less English. About 75 percent had not completed high school and about 65 percent had limited or no English proficiency, according to a report by the Pew Hispanic Center in Washington.

Immigration drove wages ever lower as large employers turned to Latino labor, recruiting through family relationships. It altered the atmosphere inside the plant, increasing racial tensions between the African American and Latino workers who typically reported to white supervisors. Resentment grew as Latinos, many hard-working and driven to succeed, began to be promoted. Blacks began to fear for their jobs as they watched Latinos slowly replace black workers, initially through natural attrition, unavoidable with the plant's consistent average 100 percent turnover every year. Later, it appeared to become part of a ploy by management to reduce wages and water down union organizing.

White managers followed the instructions of national union-busting consultants hired by Smithfield, controlling both groups by telling black workers they would be replaced with lower-paid Latinos while threatening Latinos with deportation.[24] Latino workers were, initially at least, easier to scare and control than blacks. They were fleeing poverty far worse than that surrounding the plant and were sending money to family across the border. Some had suffered horrifically—financially and physically—to cross the border, and were terrified of being sent back. Despite a long legacy of collective action, most wanted to remain under the radar, focused on building a new life.

Uneasiness among blacks grew to distrust and hate. Company management used those emotions to their advantage to keep the workers from joining forces to demand better work conditions and wages. It was a classic divide and conquer. "Smithfield keeps black and Latino employees virtually separated in the plant, with the black workers on the kill floor and the Latinos in the cut and conversion departments," Buffkin said. "The word was that black workers were going to be replaced with Latino workers because blacks were more favorable toward unions."[25]

---

Fueling the fear on the part of black workers was a series of firings of black union supporters in retaliation for their efforts leading up to the 1994 election. Three weeks before the election, supervisors had ordered Chris Council, an African American union supporter, to take a stack of six hundred anti-union fliers and hand them out to the kill floor employees as they clocked out. Council did so, but muttered to the workers that they should throw them in the trash.[26] Leading up to the vote, Council's supervisors quizzed him repeatedly about his union sympathies. "Come on, you can talk to me," one cajoled. On the day of the election, Council's supervisor ordered him to stamp the hogs with an ink stamper as they flew by on the line. After he stamped fifty or more hogs, he saw the words "Vote No." He refused to continue.

After the election, supervisors waited for an opportunity to retaliate.[27] Council had injured his shoulder during horseplay on the line opening hogs. For fun, the young men on the line would sometimes have a contest to see who could open every hog on the line, instead of every fourth hog, as plant policy required. On November 4, Council turned in a doctor's note excusing him from work. Council's supervisor accused him of forging the note, then reviewed with him a series of disciplinary actions including splashing water on the line, not wearing proper safety equipment, and horseplay.[28] On November 5, Council was called into his supervisor's office, where he was fired. There he saw a copy of a videotape of himself onstage with Rev. Jesse Jackson at the rally before the election.[29] Supervisors said that it was for cause. Council believed it was because he brazenly sat on the stage with Jackson.

Fred McDonald, also a hog opener, had gotten more than a hundred union authorization cards signed before the election. He openly handbilled, went to union meetings, and attended the Jackson rally. The week before the August election, a supervisor confronted him. "Why do you all guys want a union? The union can't do anything for you but cause trouble between the workers and the company."[30]

On December 22, supervisors gave McDonald a list of infractions, including walking off the line to use the bathroom and cutting down a hog off the line. McDonald protested that he was following established protocol, namely, that if a hog didn't have its belly facing him and couldn't be turned around, he was supposed to cut it down. But supervisors insisted McDonald had burst a hog, causing production losses. When a hog opener accidentally cuts into the guts of a hog, the USDA stops the line until the mess is cleaned up. It was a common occurrence in the plant even for a skilled knife man like McDonald. That day, despite his protestations, McDonald was fired. He was convinced it was for union activities.

Since April of 1993, Larry Charles Jones had worked in the loin-and-bone department, cutting the bones out of sirloins. He had gotten about fifteen workers to sign union cards, passed out handbills at the front gate maybe forty or fifty times, and had worn "Vote Yes" stickers and a union T-shirt. Before the vote, supervisors confronted him, accusing him of tearing company literature off the wall, asking why he wore a union T-shirt, and telling him the union would cause the plant to shut down. One supervisor gave him a "Vote No" sticker for his helmet.

By January of 1995, rumors were spreading that the company was using its complicated and inconsistent absenteeism policies—based on a point system irregularly enforced—to fire union supporters. Jones wasn't worried; he had just gotten a bonus check in October for good attendance. Then, on January 25, Jones was told that he had missed too many days, and he was fired. Jones knew it wasn't true. He knew exactly why he'd been fired.[31] He was black and he was a union supporter.

Council, McDonald, and Jones were all African American. All fell victim to company retaliation for union support. All became the fodder for whispers that spread through the plant that the company was intentionally taking advantage of the growing supply of Latino immigrant labor pouring across the border, heading for North Carolina.

## CHAPTER 7

# The Company Woman; Climbing
# the Ladder Has Its Costs

**D**uring the first union vote in 1994, Sherri Buffkin kept her head down, intent on her rapid climb up the corporate ladder. After the vote, she continued toeing the company line every step of the way, doing her part to make sure the UFCW gained no foothold.

Smart, hard-working, and ambitious, the petite blonde had moved from hourly box hanger to crew leader. By 1994, she was a supervisor over the warehouse and laundry. By the end of 1998, she had moved up to production support manager in charge of purchasing, supervising about seventy-five employees in the box room and two large laundries, as well as overseeing the janitorial staff.[1]

It was a remarkable climb for a woman in what was very much a man's world. In addition, for three consecutive years she had received the highest-percentage wage increase at the plant, another remarkable achievement. She and her husband Davie were saving for a house, hoping to move out of their single-wide trailer about five miles from the plant. Smithfield was delivering on the promises of boundless opportunity it offered when it opened. Thanks to the company, she was better off than she had ever dreamed she'd be while growing up poor in Bladen County.

As a manager, she quickly learned to surround herself with loyal friends, first hiring her husband's good friend Billy Jackson to work for her in the laundry and then his wife Susie, who was "at Melvin's flipping burgers" at the time. The two had gone to high school together.[2] In April of 1996, Susie Jackson started as a purchasing clerk, before reporting directly to Buffkin in the laundry that September. There she took phone messages, helped hand out smocks and equipment, and worked on restructuring procedures and streamlining paperwork on the first shift. The two quickly became insep-arable. They had lunch together every day and often went out to dinner, sometimes bringing their husbands, with company vendors footing the bill. They also swapped child care, since Jackson's two boys—then 2 and 10—were friends with Buffkin's 8-year-old daughter. Sometimes they celebrated their children's birthdays together and went to NASCAR races. They talked about everything, both personal and work-related.

That included the unwanted advances that Buffkin said she was getting from general manager Jere Null. Buffkin says they had consensual sex one time in March of 1995, before Null became her supervisor.[3] The next day, Buffkin says she told Null it had been a mistake and could never happen again, but she says she told Susie Jackson he wouldn't stop pursuing her.[4] At the time, Null was on a fast track at Smithfield Foods. Within two years of being hired in 1992 as a financial analyst in Tar Heel, he was promoted to comptroller at Smithfield headquarters in Virginia.[5] Then, in July 1995, Null became vice president and general manager of the Tar Heel plant, where he directly supervised Buffkin. While her initial (alleged) sexual encounter with Null may have helped Buffkin's career, she later said that after Null took over running the plant, her life became a nightmare.[6] She said that Null told other employees that he was sleeping with her and warned her to stay away from another male employee.[7] She said that she would show Null's text messages to Susie Jackson and they'd sit and laugh.[8] In photographs taken at the time, Null had short, dark hair, his round face clean shaven, and was wearing a tie and a starched shirt. "I showed Susie every time he'd page me. He would use the code 90210 because I had told him I watched that TV show. That meant the reason he was paging [was] something personal," Buffkin said.[9]

As with many women who earn fast rewards in a man's world, tongues were wagging about Buffkin. The rumor mill had Null and Buffkin in a relationship, and their flirting didn't help quell the talk that Buffkin had slept her way up the ladder. Other stories painted her as an aggressive manipulator who sabotaged the work of others, then took credit for saving the day.[10] Whether the rumors were fair or true, Buffkin received excellent written performance reviews and top raises. There was no question co-workers were jealous of her successes. In particular, Larry Johnson seemed jealous of Buffkin's relationship with Null and would razz her, telling her he hated blondes and that—just because she was blonde—he'd make sure she wound up working on the kill floor. In her view she and Johnson were in a constant rivalry.[11] (Johnson has declined to comment.)

Bolstering her claim that Null was harassing her, records show that by May of 1997, Buffkin had asked to report to the vice president of purchasing at headquarters for the portion of her job that involved making purchases for the plant, though she still reported to Null regarding other business matters.[12]

---

Aside from her troubles with her supervisor and the jealousy of co-workers, Buffkin had her hands full managing the laundry. It had always been a center of union activism because nearly every employee had to travel through it. It was also literally a crossroads, on the second floor of the plant, ringed by a hallway that led to the flights of stairs down onto the plant floor. During shift changes, thousands of workers waited in line to collect clean equipment and smocks and turn in their dirty laundry. Tempers sometimes raged. Language was forceful and off-color.

Margo McMillan, who drove in every day from Wilmington, was a laundry crew leader on second shift reporting to laundry supervisor Billy Jackson, who reported to Buffkin. McMillan had been with the plant since September of 1994. Six women ran both laundries: the conversion laundry that serviced the meat packagers and another called "the old laundry" that serviced the cut and kill floors. McMillan, a freckle-faced African American, handled orientation for new hires, answered the phones, handed out supplies, and did monthly inventory. Gloves and smocks were needed daily, sometimes more than once. McMillan, unbeknownst to Buffkin, had signed a union card she was handed in the parking lot as she came to work. She also directed new hires to attend mandatory anti-union meetings, which showed films of violence that had occurred at other plants during strikes.

As in other parts of the plant, the language was colorful and racist. A co-worker hated Mexicans and called them wetbacks, sticks, dumb butts, and motherfuckers—to their faces. McMillan reported the employee and Buffkin wrote the woman up for cursing. Another woman—who was not written up—called the Mexicans stupid asses and complained about the motherfuckers. A third confronted a supervisor, calling him a "big fat motherfucker," but was not fired.

McMillan had a reputation for being fair but firm. She was well known and well-liked throughout the plant. So was her co-worker Ada Perry, a white woman in her late fifties, who had a running riff going with the workers who lined up, impatient for their smocks. Perry, whose nickname was Grandma, had worked at the plant since December of 1993. Perry was a talker and was known for "picking," or teasing the production workers as they came through. "If they come up to me and say, 'I'm coming across that counter,' I say, 'Come on, baby, if you know it.' All the women would say, 'Grandma got you.' They might try to take something back away from me, and I'd say, 'You don't need to come across that counter.'" Sometimes the men — nearly all in their thirties — would lean over the counter and say something like, "Hey Grandma, you need a stud like me." Perry would retort, "When you grow up, then you'll be a stud."

Perry made no distinction between line workers and bosses. She'd tell them "just because you have a white hat ain't no sign you can't take the same punishment." During the rushes, if smocks on the counter ran out and she'd have to start pulling them out of the laundry bags, she would throw smocks to workers to speed things up, just as the other laundry workers did.

She sometimes asked for help if things got too unruly, especially when workers snatched smocks out of her hand instead of waiting their turn. One time, McMillan was doing paperwork when Ada interrupted to angrily tell her about a small-framed black woman who had grabbed a smock and headed out onto the cut floor. McMillan went after her, and the woman told her to "get the fuck out of my face." McMillan reported it, but nothing was done. One of McMillan's employees had cursed her out, calling her a bitch, a motherfucker, and a son of a bitch. Despite reporting the insubordination, McMillan said, nothing was done. Yet in May of 1997, McMillan was issued a warning for wearing a necklace to work. She felt the inequities in discipline were unfair.

Most of Buffkin's employees — about 80 percent — were black, and race continued to unite and divide at the plant. "I've heard them call each other nigger," Buffkin recalled, but said that she didn't think it was appropriate between races. Racial slurs rarely resulted in termination or even reprimand. In one case, in October of 1996, an African American employee told her black supervisor, "I'm tired of you niggers telling me what to do." In another case, in September of 1996, production was slowed on the line when one white woman called another white woman a "nigger lover." Another man said that an African American crew leader was a "damn nigger who thinks everybody owes him something." Some people commonly used the term "wetbacks" to refer to their Mexican co-workers. In one case, a worker told a co-worker to go back to Mexico. None of these instances resulted in disciplinary action.

Patsy Lendon, who had seen Lawanna Johnson punished for her pro-union activities, was a Fat-O-Meter reader, who recorded the lean-to-fat ratio of the hogs on the kill floor. She actively stumped for the union during the 1994 campaign and would again when the union got another vote scheduled for 1997. The 48-year-old black activist talked up the union in the cafeteria, got union cards signed before her shift, and wore union T-shirts to work. On June 25, 1997, on break from her job on the second shift, Lendon went to the Livestock break room, where everyone was also African American. She wore a yellow hard hat bearing her nametag as she handed out campaign flyers and union cards. "Are you going to vote for the union?" Lendon asked co-worker Pamela Williams. "That's none of your business," Williams retorted. "That's what's wrong with all you niggers—you always worried about yourself," Lendon snapped back. As she collected the signed cards, supervisor Lonnie Galloway told her, "You need to get your damn ass out of my department and leave my people alone." "That's a dumb-ass nigger. Don't pay him no mind," Lendon sassed. The group laughed, then began teasing supervisor Shawn Troy about his white hat, a color worn only by the predominantly white management. "I earned my white hat. I been here like four and a half years," Troy protested. "Well then, you ain't scared to sign a union card. You ain't scared of the white man, are you?" Lendon said. "No," Troy said. Lendon gave him a union card. "I worked hard for my white hat, too," Galloway said. "Yeah, kissing a white man's ass."[13] The group laughed. No one protested at the time. But on July 14, 1997, Patsy Lendon was terminated for using racial slurs. "It's just common language for black people to talk to black people in that manner," Lendon said later. "The word 'nigger' is just slang language that blacks commonly use, not a racial slur. Not for one minute do I believe I was saying anything offensive to anybody who was black."[14]

Derogatory racial slurs were common. White management and supervisors used them. Workers sometimes used them with supervisors, though mainly reserving the words for each other. Still, the irony of an all-white management team using accusations of racism to fire black union activists left the activists furious at the double standard and lack of uniform enforcement.

Obviously, union workers were being held to a different standard, ostensibly fired for cause when in reality they were being fired because of their union support and as a warning to other would-be union supporters.

---

Meanwhile, though union complaints against Smithfield—including the reasons for Lendon's termination—were piling up for investigators at the NLRB, nothing had been corrected and no penalties imposed upon Smithfield managers. By 1997, allegations included more than half a dozen supervisors illegally interrogating employees about their union sympathies.

In early July—not willing to wait any longer for the NLRB to enforce labor law—the union petitioned the NLRB for a new election, and a new vote was scheduled for August 21 and 22, 1997. On August 13 or 14, 1997, one week before the union vote, someone spray-painted "Nigger go home" on the union trailer in Tar Heel, a public display of the racial tensions inside and outside the plant. Some believed it was directed at civil rights leader Jesse Jackson, who was scheduled to speak August 14 in the auditorium of the Bladen County Community College.

Just as they had in 1994—but with more vitriol—local black ministers lined up in favor of the union, and white business owners lined up with Smithfield. Ministers from more than a dozen black churches handed out leaflets at the plant gate.[15] Company executives fired back. "It would not surprise me if the union had paid those people off," Sherman Gilliard III, director of human resources for Smithfield Foods, Inc. told the local paper, perhaps a reference to the well-worn idea of union agitators stirring up a group of people who would have been content otherwise.[16]

Members of the Lumberton Ministerial Alliance, a group of about thirty mostly black religious leaders, publicly defended themselves, saying the union had not paid churches or church leaders for their support. In their view, Gilliard was trying to inject race into the union drive. "He knows that no union has brought any money to these churches," Otis Pelham, pastor at the Sandy Grove Baptist Church in Lumberton, told the *Fayetteville Observer*. Some churches did allow union organizers to speak during Sunday services and hand out leaflets, though.[17] Smithfield's Gilliard condemned the alliance and said that most church leaders wouldn't return his phone calls. "They've formed an opinion and they don't want to talk to anybody," he said.

"That attitude tells me that some of the churches have been infected with the union." He said that he had even invited Pelham to visit the plant and pray with workers. Pelham declined. Then Pelham invited Gilliard to speak to his congregation. Gilliard didn't respond.[18]

Meanwhile, local white business owners began to register their opposition to the UFCW. "I never have seen any good that a union has done especially as it relates to community support and as it relates to other industries coming into the community," said Mac Campbell, owner of Campbell Oil in Elizabethtown. Campbell was a member of Concerned Citizens for a Union-Free Bladen County, which had scheduled an anti-union meeting at The Barn in Elizabethtown on August 14. Outside Underwood's General Store on Highway 87 in Tar Heel, a sign read "Employees Vote 'No,'" urging workers to vote against union representation.[19]

Smithfield workers were also speaking out, asking for better working conditions, job security, a guaranteed number of hours per week, and hourly pay comparable to what workers earned at Smithfield's other — unionized — plants. Workers at the other plants made a base wage of $8.45 to $10.03 an hour, while workers at Tar Heel earned a base wage of $7.40. One worker told the newspaper she had been yelled at by supervisors and wasn't allowed to go to the bathroom.

Rayshawn Ward, the popular 21-year-old who wore his hair in corn-rows, galvanized workers on the line. Ward was happy-go-lucky, quick-moving, and impulsive, with a big smile. People just liked him. He had grown up in Miami, but once he visited his godparents in St. Pauls, near the plant, he wouldn't go back. "I liked it," he recalled. "My dad would call me and say, 'Now don't let them boys hang you down there.' But I love it here." Ward went straight to work at Smithfield Packing when he was just 18.[20]

"I worked on the kill floor hanging hogs. I could hang 32,000 hogs a day," Ward said, though he was a little guy, five foot seven, weighing in at only 110 pounds. "I really enjoyed it. They'd come in live after they shocked them, through the scald tub, then after they leave out of dehair, they come to the gam table." He loved the nonstop moving line. "You got to twirl them. Every hog hits that table. The hog turner turns it to the heel string cutter, then the hanger. I would alternate from hanger to heel string cutter. Hogs coming three and four out of the scald tub at a time. Water splashing everywhere, hot scalding. We'd hang 'em and then they'd go on down the line through the fire, to the shavers, then they'd start cutting the head off."[21] Ward said that his nickname was "Young Blood," but some called him "the Union Man."

When the union reps first showed up at his house for what they called a home call, they told Ward: "We heard you had the biggest mouth at Smithfield. We heard you're the craziest man in the plant of four thousand people and everybody knows you." It was a good match. The union could parlay his popularity into a following, and Ward loved the attention. More important, even though he liked his job he was sick and tired of how the workers were treated at the plant. "I remember a guy had a heart attack, they pushed the guy off the line and kept running the hogs. . . . They wouldn't let women use the bathroom and fired women for being pregnant."[22] Ward agreed to come to a union meeting at the Holiday Inn in Lumberton. "They had drinks and sodas and everybody wanted to meet me. After a month or so I started working with them."

The week before the election, Ward single-handedly signed up a hundred people. He passed out union cards in the cafeteria and handbills in front of the plant. He wore union T-shirts to work. Ward lived in Bladen County with his wife, three-year-old child, and a new baby. He sang in the choir and ushered at Bryan Swamp Baptist Church. "I was playing that double dare stuff. I'd walk in and say, 'It's Union Time!' and Larry Johnson, he'd just look at me. He came out of a union plant. In 1996, I went all in. Union shirts, union stickers on my hat. I didn't care." Ward recalled being told that he was in the South now and the South didn't believe in unions. Supervisors threatened to fire him. "I said, 'I don't care,' and they didn't fire me. They just said I was crazy. I was young. I liked to be loud."[23] Ward said that he never feared Larry Johnson and that the plant manager never confronted him about his union activity. "He had that sneaky grin. Clean cut, always smiling. Dressed nice. Just walked around the plant looking pretty."[24]

Ward took on anyone, any time, frequently challenging managers at antiunion meetings. When they showed outdated tapes of union fights at other plants, he pointed it out. When they compared the Tar Heel plant to smaller plants, he pointed it out. "You don't want to shut up, do you?" one supervisor asked him. Another told him he "was really kind of like rude in the meetings."[25] When labor attorney Bill Barrett, Larry Johnson, and Jere Null told a group of workers they were going to hand out raises and make the plant better, Ward stood up and said they should be fired, that they weren't worth a damn, that the workers needed more money and needed to be treated right.

"The whole room was hysterical. Bill Barrett was giving me this look like I'm going to get this little joker. He was making all this money off Smithfield. He didn't have to respect us workers."[26] After that, Ward wasn't allowed to go to the anti-union meetings.

Ward wasn't the only one being watched. In August, during a mandatory anti-union meeting, Billy Jackson identified his laundry employee Margo McMillan as being from up north and knowing about unions. McMillan conceded that she had belonged to unions and did know about them. On August 13, a friend from maintenance told McMillan that upper management was asking questions about what kind of person she was. "You better watch yourself, because they are watching you," she said.[27] McMillan was worried and talked with both Billy Jackson and Sherri Buffkin. She had never worn union T-shirts or buttons or publicly supported the union. Buffkin assured her that her work was exemplary and that upper management knew it. She promised to smooth everything over.[28] If she intended to, however, she was unsuccessful.

The next day, McMillan was called in to a meeting with Buffkin and a representative from HR; she was told she had a bad attitude and would have to be moved to an area where she had fewer contacts with employees. Buffkin protested, telling HR that McMillan had great attendance and good clerical skills. Buffkin offered her a desk job in her department, away from other workers. McMillan was so upset that she put a petition out on the counter of the laundry vouching for her good character. The first day, she gathered two hundred signatures, but it was too late. She realized then that she should never have told Billy Jackson and Sherri Buffkin that she supported the union.

---

Meanwhile, the vote was fast approaching. Union organizers went into overtime, focusing primarily on the ever-growing Hispanic workforce. There had been fewer than 200 Latino workers before the 1994 vote; now there were 940. To improve turnout for the Jesse Jackson rally, the UFCW printed stickers, T-shirts, and posters in Spanish. The company did the same, even offering separate mandatory anti-union meetings in Spanish.

The total number of workers had doubled since the first vote, from fewer than two thousand in 1994 to four thousand in 1997.[29] UFCW brought in about ten full-time staff and fifty to sixty special-project union workers—

meatpacking workers who had taken time off to lobby their fellow employees
— in the final weeks before the election. The typically high turnover levels
in meatpacking made organizing even more challenging. With fewer than
8 percent of the 1994 workers still there, union staff often felt they were
starting over.[30]

Through all of this, both the UFCW and Smithfield were playing from the
usual rulebook for a contested union vote. The 1.7 million–member UFCW
used the same strategy in Tar Heel that it had used throughout the South.
The union staff was diverse and mirrored the workforce. They parlayed the
union fight into one of civil and human rights—relying on preachers and
churches—rather than focusing solely on collective bargaining and safety.
To counter that strategy, the company painted the union as a corrupt organization that would take in union dues, then lead strikes that would force the
company to lay people off and even shut down.[31]

On August 14, one week before the second union vote, a rally was held at
the union trailer just down the road from the plant in downtown Tar Heel.
Jesse Jackson and union representatives told the crowd of between 100 and
150 that the union would bring better working conditions and higher pay,
comparable to wages paid at Smithfield's unionized plants.[32] "We come here
today, not seeking welfare but a fair share," Jackson said during a prayer,
while the crowd held hands and bowed their heads. Later, Jackson did a
traditional call and response: "I am somebody! . . . Save the workers! Save
the families!" The audience echoed the words.[33]

Other speakers of national stature included Joseph Lowrey, president
of the Southern Christian Leadership Conference, and Linda Chavez-
Thompson, executive vice president of the AFL-CIO. The daughter of sharecroppers who had picked cotton in Texas, Chavez-Thompson was second-
generation Mexican American, and in 1995 was the first person of color to
be elected to one of the AFL-CIO's three highest offices. She was there to ac-
knowledge the growing importance of Latino workers for the union. During
the speeches, the Bladen County sheriff's department "saturated the area
with deputies," claiming that it wanted to make sure nothing happened.[34]

The union served up fried chicken and potato salad as news reporters
interviewed attendees. Margo McMillan was there, along with laundry co-
worker Ada Perry. Perry hadn't been involved in the union during the 1994
vote, but in 1997 union organizers came by her house to invite her to hear

Jesse Jackson. At the rally, she felt emboldened. "We're gonna win, honey. We got it," she told a journalist. The *Fayetteville Observer* reported her name and that other workers called her "Grandma."[35]

When the newspaper hit the streets the next day, there was hell to pay. Buffkin was called in to plant manager Larry Johnson's office along with general manager Jere Null and second-shift plant superintendent John Hall. They all knew that Johnson had a soft spot for Grandma and would always help her out if there was any trouble. They could see he was angry about the newspaper that lay on his desk. John Hall started laughing and joking with Johnson. "I guess Granny is not going to come running to you any more, is she? Her days of running to you are over." Johnson shook his head. They were indeed.

Five days later, he went to the window of the conversion laundry room and told Ada Perry they needed to have a talk outside, alone. She went out the back door and met him outside.

"You hurt me so bad. I can't believe it," Johnson said.

"What are you talking about?" Perry replied.

"You're a union organizer?"

"No, Larry, I just went to see Jesse Jackson."

"You're all over the front page. Are you going to vote yes?"

"That's everybody's decision, like Democrat or Republican," Perry said.

"Will you please vote no? Will you please just tell everybody to vote no when they come up here to the window?"

"Why me? Why are you asking me out of all these people down here?" Perry asked.

"Because this whole plant looks up to you," Johnson told her.

"Well, why ain't I getting more money if I'm so important?"

"Go down and see Jere. He'll make it good."

Perry didn't go see Null. Instead, she agreed to be a union observer on the day of the vote.

Buffkin tried to stay out of the fray. Her initial policy with those loyal to her was live and let live. "When Jesse Jackson and they all met down at the trailer," Buffkin said later, "I did not participate in that and I wouldn't today. That's not me. I have nothing to do with them. I knew Ada and Margo went and several others. I did not care. It weren't my business. [Smithfield] had built me my own office where there was none. I had my own big-ass oak desk and chair and three secretaries. I was the third-ranking person in the entire facility. You didn't tell me what to do."[36] Buffkin had worked with Perry and McMillan for years. She knew them well, knew their strengths and weaknesses. She could protect them, or she could report their union activities. She could discipline them, or she could fire them. But pressure was building.

---

In the days before the election, production slowed to a crawl so that every worker could attend large, mandatory, plant-wide meetings known as "25th Hour Speeches." A series of smaller meetings culminated with huge compulsory meetings for all shifts the last day before the vote. Supervisors told workers to report to the box room in shifts. On August 19, 1997, there were at least two meetings, with half the plant going to the first and half to the second. Conversion and cut floor went first, then kill floor and casing. Buffkin was in charge of setting up the box room for the big event. The warehouse was emptied out. Supplies and boxes were loaded into trucks. Chairs were set up. Windows were blocked to keep out the light. "It was packed," Buffkin said. "I mean, people were sitting on what few boxes we had left in there. Every chair was full. The aisles were full."[37]

A sound system, TV screen, and teleprompters had been installed with a large podium. Hispanic workers gathered in a nearby room with a translator. General manager Jere Null spoke to workers for forty or forty-five minutes. Null stood on a platform behind the podium, but as he spoke he moved from left to right, away from the podium, surveying the crowd. Behind him was a screen with his picture on it. According to workers who attended the meeting, he echoed the same claims his predecessor, Henry Morris, had made in 1994.

"Before this plant opened, all there was here was tobacco fields and peanut fields," Null told them, adding that if the plant closed, there would be nothing again. "We brought jobs, and we are not going to let the union in without a fight." He told the workers about nine other plants he said had been closed down by unions. "I don't want to lose [my job] and I'm sure you don't want to lose yours."[38] A worker who was there recalled, "That's when everybody starting moaning that they can't afford to lose their houses." And they were sure they would if they lost their jobs, or if the union was voted in and called for a strike.

Null walked away from the podium, then came back talking about unions. Unions mean losing jobs and losing cars and losing houses, losing wages. He shook his finger as he reminded workers about what had happened at the other plants. "If you sign union cards and they go out on strike, the company is not going to budge. We will replace you. You won't get unemployment and you are not getting welfare. . . . If the union is voted in, [the raise planned for this fall] will automatically stop. It's out of my hands. My hands are tied. It would have to be negotiated by the union in their contract."

If the union won and called a strike, workers would lose health insurance, he told his listeners. "There will be no negotiations because all they want to do is come in and take money out of your pockets. You shouldn't let them because they want to get rich off of you. We took you out of the fields as being slaves, out of the bean fields and potato fields and you don't want to be out there on welfare and you know you can't live off that strike pay, which is thirty bucks a week. Before we let that union come in, we'll close this plant down, start everybody off at minimum wage and reopen it."[39] Smithfield had saved in the range of tens of millions of dollars in wages since the plant opened by keeping the union out. To organizers the accusations that the union wanted to get rich off the workers was laughable. Getting rich by paying low wages was one of the biggest reasons the plant had been built in North Carolina.

Along with the mandatory anti-union meetings during the last few days leading up to the vote, management showed videos of violence, arrests, and strikes in other union-organizing campaigns in the meatpacking industry. One worker was beaten when he crossed the picket line. A video showed a bullet hole over a baby's crib where union workers shot into the home of a scab, a non-union worker who crossed the picket line. A garage was burned down. A worker was arrested and led off in handcuffs. Every video ended with the words: Vote No!

Workers on both sides agreed that Null's last-minute meetings — four the Wednesday before the vote began, with more than one thousand workers — may have been a turning point. During at least one of them, Null was booed when he began and cheered when he finished.[40]

When the last meeting ended and Buffkin was clearing things away in the box room, Null spoke to her privately. "He told me he wanted . . . Ada Perry terminated, but he told me to wait four or five months till all of this blew over. . . . He then proceeded to kindly reprimand me, to tell me that my laundry was just a hothouse of union support."[41]

Bill Barrett had another target and he wanted immediate action. Barrett had just left an anti-union meeting, where Margo McMillan's name had come up repeatedly. Barrett looked intently at Buffkin. "Fire the bitch," he said. "I'll beat anything she or they throw at me in Court."[42] Buffkin was in a panic. McMillan was a good worker and a friend, with a family to support. There had been few write-ups, all minor, in the years she had worked there. She told Barrett she couldn't do it, because there were no disciplinary actions in her file and there were no grounds. John Hall, the second-shift plant superintendent, said that he would take care of it. Jere Null came out of his office and said, "Do it."

Buffkin begged Hall. "She's got family problems. She's got kids. She's got bills to pay. I'll be responsible for her. I'll take care of her and you don't have to worry about her being in the laundry or being in contact with so many people." Hall agreed. That afternoon, Hall called McMillan into his office with Buffkin and told her she was being moved to a solitary job handling labels. Margo was desperate, asking repeatedly why she was being moved. Hall told her it had been brought to his attention that she had a bad attitude and had to be removed from the general population to a controlled area.

McMillan was desperate to keep her job. She felt Jackson and Buffkin were both lying to her, telling her to her face she was a good worker but reporting to management that she was not. She took the petition to Larry Johnson, the one with two hundred signatures of co-workers vouching for her good work and good character. Johnson responded by telling her she was hated by the Hispanics. As she left, she overheard Buffkin telling supervisor John Hall, "That motherfucker was up in Larry's face with a paper." McMillan knew she meant her.

When Jere Null heard that McMillan had been offered another job under Buffkin, he nixed it. "No, no," he said. "She cannot have that job." When Hall told McMillan that the offer of another job had been rescinded, McMillan broke down, sobbing hysterically. "She begged. She was crying. She was, yeah, was hysterical. She was begging. Told him she would do anything," Buffkin recalled. Hall sent McMillan home and told her he would let her know. Three days later, on August 21, when McMillan came to observe the election for the union, the paycheck clerk told her she was no longer employed.

## CHAPTER 8

# The Second Union Vote; More Firings

Thursday, August 21, 1997. The first day of the two-day vote had arrived. When the second shift started midafternoon, workers would begin to cast their ballots. First-shift workers would vote the next day, then the votes would be tallied.

Union reps, along with Jesse Jackson, got to the plant around 5:30 a.m. to handbill first-shift workers, urging a "yes" vote as they drove into the parking lot. About seven to eight hundred cars drove past between 5:30 and 7:30 a.m., some swerving to pick up or avoid getting handbills, some speeding up — the usual rhythm. Some employees drove alone; others carpooled. Chad Young, the young African American leader for the UFCW union campaign, was there. Young, who had moved from Louisville in March, served as the assistant to the regional director out of the Atlanta Office.[1] Hispanic union organizers were betting on a 50/50 split, while African American organizers bet on a 60/40 vote in favor of the union.

Head of security Danny Priest was also there, with about four or five deputies. Jere Null and Larry Johnson also showed up.[2] "Chad was in the back of the group and [Rev. Jesse] Jackson came out and introduced himself to Larry Johnson and me," Null recalled. "He said this was an impressive plant and he had worked in a packinghouse in South Carolina years before. He asked us if we still scalded hogs the same way. . . . I remember telling him under other circumstances, I would love to take him in, but I didn't think it was appropriate. I told him they weren't allowed to be on the property."[3] Then Null and Chad Young began to argue over whether the union had exceeded the allowed two days of handbilling per week. Null said that they had. "Then arrest us," Young said. "Because we're not going to leave."

"This is pretty pathetic, bringing Jesse Jackson in on the last day," Null said. "This is an act of desperation."[4] Company supporter Anthony Forrest, a large black man who sharpened knives in the plant knife shop, chimed in: "Yeah, you guys are going to get your ass whipped tomorrow." Young lost his cool. "Why don't you shut up, you house nigger," he snarled at Forrest—or something like that. Some said that he actually called him an Uncle Tom.[5] Forrest bristled. Larry Johnson, who had heard the exchange, lifted his hand and said, "Anthony, it's not worth it."

The two backed off, and everyone went back to handbilling. More deputies arrived, parking in the grass along the curving driveway that ran up to the guard shack, which sat about a hundred feet off Highway 87. By the time the shift was in full swing, eight to ten uniformed guards were there. Some deputies wore flak jackets and some held shotguns.[6]

Null invited Young to the vote count the next afternoon. "I want to make sure you're there for a real ass-whipping," the general manager said. "We're going to beat you by at least two to one. And, we've got something special in mind for you, so I want to make sure you're there." "I'll be there," Young said.

The union handbillers went back to the Holiday Inn to regroup for the second shift, while NLRB representatives inspected the voting room, preparing to oversee the election as usual. The vote would be in the smaller second-floor cafeteria, called the nonsmoking cafeteria, a room about 60 by 35 feet. Five tables were set up along the wall across from the windows, with space behind them for NLRB officials. Chairs were set up for union and company observers.

Shortly after noon, union handbillers came back, parked across the highway, and walked to the guard shack. Heat radiated from the road in the near 100 degree weather. Most organizers wore cut-off clothes or sleeveless T-shirts and shorts. Inside, union observers were arriving. When Jere Null saw that laundry worker Ada Perry had volunteered, "he turned real red and left," she recalled, because he thought she was going to go along with the company as Larry Johnson had asked.

As the lines were shut down so employees could leave to vote, neither side was allowed to speak or campaign in any fashion. Each eligible voter made an X on a paper ballot for yes or no and slipped it into a ballot box. As voting began, Ada Perry checked people in, making sure they were on the list approved by both sides and were eligible. Supervisors and new employees were not permitted to vote. But Perry found people who weren't on the list and suspected the company of sending supervisors, new employees, and out-siders in to vote against the union. "A black supervisor came in there, and I said, 'No, no, no,' and told him he couldn't vote. He was from shipping. He was there putting on his freezer suit and his white supervisor hat on."

It wasn't the only thing that made Perry suspicious. A little after 9 p.m., when the kill floor came to vote, the lights suddenly went out. It was pitch black, and people in the room started to scream. Some yelled "Get the box!" and "Watch the ballot box!" meaning to make sure it wasn't tam-pered with. Some turned on flashlights. When the lights came on some five minutes later, a fired worker—who had been rehired as an anti-union consultant—was hovering over the box, which had been left unattended by NLRB agents during the blackout.[7] The union, however, was never able to prove tampering.

The next day, August 22, as voting continued, the popular and sassy Ray-shawn Ward was to be a union observer. He was excited. He normally would have driven the thirty-five minutes from his home in Bladenboro, swiped his ID badge, and checked in by 5:00 or 5:15 a.m. This day, however, he was officially working on union time and was picked up at home. "I had a chauffeur. They took me to breakfast. I felt like Magic Johnson winning a championship. It was a good day."[8] He felt proud as he headed into the locker room, changed clothes, and attended a meeting with union supporters and company managers to discuss rules of who would vote and how. He was uncharacteristically serious as organizers met with NLRB representatives to discuss proper conduct.

NLRB officials carried ballot boxes upstairs to the nonsmoking cafeteria where the five long cafeteria tables were again set up on painted concrete floors along the cinder-block walls. The officials stood across the tables from the union and company observers, who were placed in alternating positions, one after the other. Ward was paired with a company representative. To-gether they walked through the plant from line to line, holding up a sign for workers saying, "It's time to vote." The lines would shut down, workers would vote, then head back to the line.

Sherri Buffkin had been nervous all day. The atmosphere was tense. Along with all salaried employees, she had been ordered to the cafeteria during the vote count. That violated NLRB rules that only small and equal numbers of representatives from each side could be present. "They didn't care," Buffkin recalled.

Outside, sheriff's deputies carrying shotguns patrolled the parking lot. "I was terrified," one worker said later, adding that it reminded him of stories about civil rights protests in Birmingham, Alabama. "I had never been involved in anything like that."

Buffkin had a feeling something bad was going to happen. She tried to shake it off as she headed for the cafeteria. Managers Jere Null and Larry Johnson stood under the drop ceilings and fluorescent lights. Danny Priest, head of security, stood nearby wearing a white shirt, his deputy's badge clipped onto his belt. A gun was hidden in an ankle holster beneath his pant leg, handcuffs were stuck into his belt, and he wore a bulletproof vest. A small canister of mace was in his pants pocket.

Chad Young arrived about 3:20 p.m., along with the union's regional leadership. Priest and Null met them at the front door to the plant. "You can have ten representatives at the vote count," Null told them. "How many are the company going to have?" Young asked. "Ten," Null said. Ten union representatives went in, while about seven or eight stayed outside. But inside the cafeteria, there were already between twenty-five and fifty company people.

The NLRB agents began the count. No one from the union or the company was allowed to touch the table as agents dumped the ballots out in front of the observers. Young went up to Tony Scott, head of the NLRB proceedings, and asked him why the numbers of company and union reps were unequal. Scott, in turn, asked Null, who responded by winking at Young and saying, "It's my plant and we'll have as many as we want." Young told Scott he was going to bring in the rest of the union reps. Outside, where Young had gone to have a cigarette, Danny Priest found him and told him to come quickly, that the room was filling up fast.

When Young returned, the atmosphere had become more tense. As the official count began, there was a deadly silence. "Nobody was saying nothing," one observer later said. Young hung close to Tony Scott as both company and union observers were being pushed forward against the table by the growing crowd. Scott told them to calm down.[9] "It was getting hostile," Ward recalled. "Stop pushing. Everyone step back or I'll halt the election,"

Scott said. The crowd stepped back and the count resumed. NLRB counters called out the votes. "Yes. No." Union and company observers tallied the votes in their own notebooks.

As the "No" stack grew, company managers pushed and elbowed to see. By the time the first count was completed, union observers complained that they were being slapped on the backs of their necks by company employees. Some felt knees digging into their backs. Sprays of moisture hit the backs of their heads and necks; they thought it was spit. "You need to do something about what's going on here," Young told Johnson and Null. "There are only supposed to be ten representatives from each side."

It's not clear whether Tony Scott was aware of the encounters, but nothing was done to clear the room. As the second and final vote concluded, a victory yell built and the cafeteria doors swung open with a roar as a crowd of managers, supervisors, and superintendents, wearing their signature white helmets and smocks, surged in. Wearing "brand new frocks, brand new hardhats, they came in and assembled around the perimeter of the vote room," Young later testified.[10] "I believe they were hired security—they were big. They were all over six feet, towered over me. Plus, they were wide. First, they reported to Jere Null, then they spread out and surrounded the room. It happened fast, three minutes."

The surge of people pinned both company and union observers against the tables. The newcomers began to chant "Union go home." Null stopped the chanting and said, "Wait for my signal." The room grew quiet.

Finally, NLRB agent Tony Scott was ready to announce the final tally: 1,910 "no" to 1,107 "yes." The company had won. Ward was standing next to Bill Barrett. "He was laughing like hell at me," Ward recalled.[11]

Null gave the signal and company supporters began to yell: "Union scum, go home! Union scum, go home!" All hell broke loose. The crowd began to move in toward the tables, yelling and screaming. "You've got one hundred rednecks in a small room," Ward said.[12] "Some of them came off the leash. I looked at [the union organizers] and I said, 'Y'all boys better put on your fighting shoes cause we're going to ball here.'"

Organizer Chad Young told Null the union would be back, yelling, "We're not going anywhere!" That's when a large man closed the door and the crowd surrounded the union representatives, pushing, shoving, and throwing

Jere Null, the general manager of the Tar Heel plant, accepts congratulations from employees after workers at the plant rejected a bid by the United Food and Commercial Workers Union to represent them. Photo by Jay Capers, *Fayetteville Observer*, August 23, 1997.

punches. "I was in the middle of the room. Somebody grabbed me from behind . . . in a choke hold," Young recalled. Another union rep tried to help, but was punched. Someone held back his arm, keeping him from signing the election tally sheet.

Then the racial slurs and chants began: "Nigger lovers go home!" "It was a riot, we're being kicked. We're being punched, we're being backed by the crowd, pushed completely out of the room," Young said later.

Sherri Buffkin jumped up onto a round Formica cafeteria table to see and to "get out of the throng," she said. "People were just screaming 'Get out, get out!'"[13] James Blount, a large, heavy-set black manager, stood on another table yelling, "Nigger go home!" Buffkin was shocked. She was used to black co-workers using the N-word in jest but not in anger. "Nigger lovers go home!" supervisors chanted at union reps. "Union scum go home! Union scum go home!" One man yelled, "We don't want any fucking union!"

Plant manager Larry Johnson turned to Anthony Forrest, the supervisor who had earlier had words with Chad Young in the parking lot, and pointed the union organizer out in the crowd. "Hey, there's that guy that called you a house nigger," Johnson said. "Now's your time to get him. I bet he's not so brave now. Go kick Chad Young's ass."

Forrest got right up in Young's face. "You're a union asshole," Forrest told him. "Get the fuck out of here. We told you we were going to whoop your ass and we did." As company men closed in on Young, another union organizer pulled him out of the cafeteria. Young heard screaming but didn't know who it was. "They were pushing us, shoving us, spitting on us, kicking us, calling us niggers, calling us union lovers, calling us everything under the sun," the organizer later reported. "We kicked your ass one time, we kicked your ass again," Smithfield managers chanted. Company employees and supervisors moved like a tide, forcing union organizers out the double doors of the cafeteria.

Then a deputy, whom Young had introduced to Jesse Jackson, suddenly appeared. "Chad, it's me. You need to get out of here." The deputy led him down the narrow hallway—now packed with people shoulder to shoulder—which became a gauntlet as union observers were groped, pushed, and shoved from both sides. Young was pushed out and down the stairs along with other representatives. A sheriff at the bottom told him to get his ass out or he'd be arrested.

Others weren't so lucky. Rayshawn Ward was caught upstairs in the hallway when he turned back into the crowd to find his wife, who also worked at the plant. He was "like a rag doll or little punching bag," one man later said.[14] "Please stop shoving," Ward protested. "We leaving the building. It ain't got to be like that." Dale "Big Country" Smith, superintendent over the cut floor, hit Ward in the back of the head so hard that he almost went down. Ward swung around. "Who just hit me?" In moments, he was completely surrounded by men in smocks and white helmets kicking him and spitting on him. "They were beating the living crap out of him," a union rep later said, calling it a "riot."[15] "I could see Jere Null just kind of with a grin on his face . . . and (a union guy) was yelling, 'Don't ever sic your dogs on us, Jere. Don't ever sic your dogs.' I've been in a lot of elections. I've never been involved in anything like that."

When Ward hit the floor, head of security Danny Priest kicked him, then sprayed him with mace and kneed him in the back. Priest handcuffed Ward, who lay writhing and moaning. "Sheriff's Department—you are under

Rayshawn Ward, who was attacked by company supporters during a near riot after workers rejected the union in 1997. *Fayetteville Observer*, September 29, 2002.

arrest!" he yelled. Johnny Rodriguez, a union representative from Dallas, was also handcuffed when he tried to help Ward.

As Priest hauled Ward down the steps and out the back entrance, newspaper camera flashes captured the young worker's face, streaked with tears and grimacing from the intense burning of the mace. Ward's wife ran to him, crying. "I don't know what they're going to do," he told her and asked her to take everything out of his pockets—keys, wallet, ID. As Chad Young yelled for deputies to stop, the deputies pushed Ward into the police car. "You're taking the wrong guys here," Young yelled. The crowd was still chanting, yelling, and screaming. Amid the violence, company employees hustled the remaining union reps off the grounds.

Police officers talked with Jere Null for about thirty minutes before taking Ward and Rodriguez to the sheriff's station. There they charged Ward with damaging the police car with his feet and striking Danny Priest. The young man faced two counts of assault, one count of property destruction, and one count of inciting a riot. Johnny Rodriguez, an employee of the UFCW, was also charged.

Union organizer John Renee Rodriguez being taken out of the Tar Heel plant after the union was rejected by a vote of 704 to 587. Photo by Jay Capers, *Fayetteville Observer*, August 23, 1997.

To avoid criminal charges, the two men signed waivers of liability in front of a judge, promising not to sue deputies for their arrest or injuries. "When I got bonded out, I went to the doctor. My head and face were burning from the mace and swelling," Ward recalled. "The mace hurt me more than the punches. I couldn't defend myself." He was furious at what he perceived as entrenched racism. "It wasn't that we couldn't eat here or sit here," as under the old Jim Crow laws, "but I couldn't voice my opinion without rednecks and white boys running us out of town because that's what they've done all their lives."[16]

When Ward came to work that Monday, August 25, he was stopped at the guard shack and escorted to a meeting with Jere Null, Larry Johnson, and Danny Priest in Johnson's office. The plant manager asked what had happened. Ward told him that he wanted a union and things got heated. "I'm just tired of this union shit," Null told Ward. "I'm ready to get my company back where it belongs." Johnson sent Ward back home and told him he would let him know whether he still had a job. The next day, at 3:00 p.m., HR called Ward at his house and told him that he was fired for "fighting." Ward believed he was singled out because of his success as a union organizer and because he had been outspoken. The union added Ward's case to its growing list of complaints to be filed with the NLRB against Smithfield.

---

A day or two after the election, Joe Luter III and Dave Barry, an executive with the UFCW, had a terse conversation on the telephone. The two men both believed they had worked out an agreement before the vote that neither side would break the law. Now that agreement appeared to have been broken. Barry complained that union workers had been roughed up by sheriff's deputies. "I told him I knew nothing about it, but I would immediately get into it," Luter recalled.

The two had spoken several times between the 1994 and 1997 elections. Both men—along with other national figures—were tracking the union battle. Other Smithfield subsidiaries had union contracts, including John Morrell, Patrick Cudahy, SK in Baltimore, Valleydale Foods in Roanoke, and Lykes in Florida, and both men wanted to keep the peace.[17]

About six months before the 1997 election, Barry had called Luter to arrange a meeting at Smithfield headquarters, then located in Norfolk, Virginia. The first meeting—described as friendly by Luter—was partly about union representation at other UFCW plants. Barry also proposed a solution for the Tar Heel plant, suggesting that Smithfield—as was allowed by federal law—simply agree to recognize the UFCW to represent the workers based on the numbers of union cards signed—rather than hold another election. "He was making assurances to me that he was not anti-business. I tried to make assurances to him that we were not anti-union," Luter recalled. The men agreed that they had different interests at times but that both believed in democracy and the people's right to decide whether to be represented by a union. So Luter told him he could not agree to union representation without an election.

The meeting wasn't entirely cordial, though. As Luter recollected, Barry pointedly told him that the union lost the 1994 vote because Smithfield had broken labor law by intimidating and threatening workers.[18] The Smithfield CEO strongly disagreed. "I told him . . . that practice might have been productive many years ago, but in the 1990s in America if you threaten and intimidate, quite frankly, I think it's counter-productive."[19] Luter assured

Barry that the policy of the company was open, free elections. "But I'm not going to push a union down someone's throat. . . . In North Carolina, people have decided over and over the union's not in their best interest." As for union-busting? Luter said, "We don't have the ability to bust the UFCW. You're a very big powerful union."

The meeting ended pleasantly, according to Luter, and the two met a second time several months later on Luter's boat in Palm Beach, Florida. The two men hammered out some detailed agreements for the August election, specifically agreeing that union organizers should not have to stand on the busy Highway 87 because they weren't allowed on company property. Luter said, "I told him I certainly didn't want someone killed on a busy highway handing out literature." He agreed they could stand on the edge of the property.

Barry again brought up company threats and intimidation of employees. The two men agreed to appoint lower-level representatives who would work together to immediately resolve complaints as they occurred leading up to the election. Luter promised to get personally involved if there was a problem.

"Obviously I wanted something in return," Luter said. He wanted a promise from Barry that if the UFCW lost the election, it would not attempt to organize the plant again for three years. They negotiated, with Barry proffering two years, but Luter held firm. According to Luter, the two powerful men finally agreed on three years. In return, Luter promised that if the union won, he would quickly agree to a contract competitive in the industry.[20]

A letter from Barry, dated June 30, 1997, put the agreement into writing. Luter sent it to Tom Ross, vice president of human resources over Smithfield Packing, and asked him to make sure the company fully complied. Though based at Smithfield headquarters in Virginia, Ross traveled to the Tar Heel facility frequently during the 1997 campaign.[21]

Luter and Ross rarely spoke, but when Barry and Luter got off the phone after the 1997 vote, Ross was the first person Luter called. Next he called general manager Jere Null in Tar Heel. Based on those conversations, Luter concluded that there had been no violence or intimidation. "I think some of our people rubbed . . . mud in their faces," Luter said. "I chastised my people. I said, 'It made you feel good for five minutes, but that's not in the company's best interest. It better not happen again.'"

Null denied Barry's allegations that workers had been roughed up, instead painting the confrontations and arrests as a spontaneous celebration that had gotten a little too exuberant. There had been a lot of shouting, he admitted, and taunting of the losers, like during an NFL football game. He did not tell Luter that he had led the cheers—that he had told his supporters when to cheer. He did not tell Luter that he had orchestrated and approved the arrests.

Once Luter had Null's report, he called Barry back. "I apologized, told him I was unhappy about it and that I chastised my people," Luter said. When Barry protested that workers had been roughed up and arrested, Luter sought to calm him down. "Dave, you're listening to your union organizers who are very upset because they lost the election. You all have spent an awful lot of money and you expected to win and now you've lost. I think you've just got a bunch of organizers down there trying to cover their asses."

According to Luter, the two agreed that they really didn't know what had happened because they weren't there. Then Luter pressed Barry on his agreement to not hold an election for three years. Barry refused. Luter was furious. "I said, 'Dave, up until now we've got a pleasant relationship.' I said, 'Don't ever call me again . . . because you've broken your word to me.'" (Barry's version of events has never been made public, and he has made no public comments regarding his conversation with Luter.)

Luter would later say that he was never involved in any details of the company's responses to the union campaign but that he had directed Tom Ross to make sure the company did not break any laws, including national labor laws.[22] If it's true that Luter knew nothing of the details of how his managers were fulfilling his publicly stated desire to keep the union out, then the distance between the executive offices and the plant floor seems to have been greater than the miles separating Smithfield, Virginia, from Tar Heel, North Carolina.

---

Meanwhile, at the plant, it was time for another sweep of union supporters. Ward wasn't the only union organizer to lose his job. Following the union vote, Sherri Buffkin was ordered to clean house, but to take her time and pick her spots. Ada Perry was a prime target. Not only had she acted as a union observer, she had gone public with her union support in the weeks leading up to the union vote. Manager Larry Johnson had taken it personally. Null had told Buffkin to wait four or five months, then fire her.

Five months after the election, an opportunity presented itself. On January 26, 1998, Buffkin got a call from Perry, a petite white woman. "The lines and the Mexican and the Black was pushing and threatening to stomp asses and all this," Perry later said. "I called Sherry, I said, 'We need somebody out there to watch that line, because we can't correct them.'"[23] Buffkin asked John Hall, supervisor of the second- and third-shift production employees, to help keep an eye on things. The next day, on January 27, Perry was working the counter at the laundry. After the workers from the cut and kill floors had been through the line, three young men in their late twenties walked up, all over six feet tall with defined body-builder muscles. Ada was bagging dirty smocks. "You all want a smock?" Ada asked. "Yes," they said. "Where's your ticket?" One put his ticket on the counter and said, "Throw it." Ada tossed the smock. The next one laid his ticket down and clapped his hand, signaling he was ready to catch his smock. Ada threw him a smock. When he reached out, teasing like he was going to take his ticket back off the counter, Perry threatened, "Uh-uh. You don't want Grandma to come across that counter." They all laughed and left. It was nothing that hadn't happened plenty of times before. It was the kind of teasing that had long endeared Perry to her co-workers.

Hall called Perry into his office. He told her the three men had filed a complaint against her for instigating a fight. Perry was five foot three and a half and weighed about 145 pounds. "Well, it's strange to me," Perry told him. "They all laughed [at the time]." Perry was fired for "threatening to do bodily harm." "I was dumbfounded. I mean it was hilarious to think that Granny was instigating a fight with somebody on the kill floor," Buffkin later said.[24]

Hall started laughing as he told Buffkin that Johnson had approved the decision. Buffkin told him it was a serious problem. "John, Granny is real involved in politics. You're going to have a problem. She's going to call [President] Clinton. She's going to call the governor." Hall said he didn't care, that he had already sent her home and told her he was going to investigate. "I don't give a damn who she calls. Maybe she can go work for the union."

On January 29, Perry was called back in for a meeting with Hall, Buffkin, and a representative from HR. Buffkin didn't say a word as Hall took charge and told Perry he had finished his investigation and she was being fired because she threatened her co-workers. Perry said she was just joking around and playing.[25] She refused to sign the disciplinary action. Buffkin signed the final termination paperwork on February 2, 1998.

---

That February, Buffkin had a lot on her plate. She and her husband Davie had just bought a new house. Their salaries together covered the mortgage and payments on two trucks. "When we moved in, we had the furniture from our trailer, hand-me-downs from the family, and that bothered me. Everything was worn out. I took out the paneling and we went to work. We could do repairs and fix things up."[26]

Salespeople, eager to gain favor with the purchasing agent, were paying for dinners, shows, and tickets to NASCAR races. Sherri often invited her good friend Susie to come along and kept her posted on the latest developments with her boss, Jere Null. Null had confided in her about problems in his marriage. "He was always complaining about her. I told him, you need to bear with her. I told him about [my daughter] Nicole and day care. I really thought we were friends. I honestly thought he was somebody I could trust and confide in, and I had on many occasions."[27]

It seems at times Buffkin thought of Null as more of a friend than a bully who sexually harassed her. Perhaps there were times she was simply in denial. After all, Buffkin was in a world beyond her dreams, a world where "people put me on a corporate jet and would fly my little country ass to Virginia to teach people in Smithfield how to do things correctly." She may have bootstrapped her way up into this world of management, but, she said, she had learned how to do things more efficiently and save more money than the professionals at headquarters.[28]

The ever-courting salesmen must have seemed like part of a dream as well. Desiring access to her multimillion-dollar budget for purchasing labels, boxes, tape, cleaning supplies, janitorial equipment, and laundry supplies, salesmen gave her gifts including basketball tickets, football tickets, NASCAR tickets, a bottle of Crown Royal, wine, clothing, chocolates, and Lone Star Steakhouse gift certificates. According to Buffkin, they paid for motel rooms in Talladega, Alabama, in April of 1998 and Daytona Beach, Florida, on February 15, 1998. Other plant supervisors knew about the Talladega trip because she was so excited that they told her to stop talking about it. "I'm leaving on a jet plane," she told them.

Other managers also got gifts, Buffkin said. Larry Johnson's "was nicer than mine. Larry got an actual dinner hamper. I got a gift basket with fruit. It was a nice stand, but they got the wicker basket, the picnic basket with the plates and utensils and all." Vendors took Null and Johnson pheasant hunting and on fishing trips, and Null got a shotgun and a civil war pistol.[29]

But there were times when Buffkin was jolted out of her dream life, playing figurative dodgeball with her boss who, she alleged, simply wouldn't leave her alone.[30] At one point, Jere Null told her "he was going to do nothing for the next four weeks but make her fall in love with him."[31] Buffkin repeatedly complained to plant manager Larry Johnson, but there was nothing more to be done. The company had complied with her requests to report directly to headquarters on all purchasing matters. But everyday concerns — especially disciplining and firing union supporters — required her to interact with Null. So she kept doing what she had to do to keep the big paycheck coming in. She kept obeying orders. After all, she didn't care a whit about the union or whether they won or lost, just so long as she could stay out of elevators with Jere Null.

---

By spring of 1998, NLRB investigators were busy taking and reviewing hundreds of affidavits from witnesses, preparing them for trial. By midsummer, even before all the charges were completed, their files filled twenty folders in five boxes. By late summer, investigators had enough evidence to charge Smithfield Packing with 286 independent violations of national labor law, section 8 (a)(1), for threats to employees and 28 violations involving employees fired for union activities, section 8 (a)(3). There were so many alleged violations that the NLRB had filed for an injunction in federal court to stop the company from firing and threatening employees and to have all fired employees rehired pending the NLRB hearing scheduled to begin in late fall of 1998. They failed.

Meanwhile, Smithfield's top managers and lawyer Bill Barrett were scrambling to meet legal deadlines. To address the most serious charges, the company would have to prove the firings were for cause, not for union activity. Margo McMillan and Ada Perry were especially worrisome for the company. Both were well-liked laundry employees with good evaluations, and both had worked for and been fired by Sherri Buffkin. Infractions were so common in the plant that most employees had numerous absentee warnings or other actions in their files, but McMillan's record was clean. She had always shown up for work. "I went to Larry Johnson and told him he was going to have a problem. There was nothing in her file," Buffkin said. Johnson told her to insert a written warning into the file of every employee of the laundry telling them to be respectful of other workers.[32] Buffkin prepared the letters, dated September 16, 1997. They warned all laundry employees that their attitudes toward other employees were less than professional and that further instances would result in disciplinary action. The letters were read aloud to each employee, witnessed by a supervisor, then placed in their files whether the workers signed them or not. With that simple action, the company had cause to fire anyone in the laundry simply by asserting that his or her attitude was unprofessional.

Attorney Bill Barrett must have known that Buffkin despised him and resented his educated manner of speaking and know-it-all attitude, but the company was on the line. They were both going to do whatever it took to keep the company safe. Or so Barrett thought. What he didn't seem to realize was that Buffkin was growing increasingly troubled by the lies she was being asked to tell about why she fired workers. And her stress over Jere Null was becoming unmanageable.[33] She was still sharing his voicemail messages with her friend Susie Jackson.[34]

In or around March of 1998, Buffkin told Larry Johnson she could not work for Null anymore because he would turn every business conversation into a personal one.[35] Then in July or August, Buffkin went back to Johnson again. This time, he agreed to try to intercede. He told her to come to him first with any business concerns.[36] But the workaround didn't help. Though she was clinging desperately to the American dream, it was slipping from her grasp.

By late August, Buffkin had to get out. She left on vacation and didn't come to work on Monday, September 7. It's not clear exactly what was going on behind the scenes, but on September 17 she was suspended for one week on charges that she gave away two cheap phones and two umbrellas that had been donated by vendors for an employee picnic. On September 23, 1998, Larry Johnson phoned Buffkin at home with Null's consent. "As of today, you no longer have a job. You have a choice, Sherri: resign due to personal matters or be fired." Buffkin remembers telling him, "You better have a damn good reason because I haven't done anything wrong."[37] He replied, "At this point, Sherri, between you and me, I no longer trust you."

Buffkin immediately drove to Smithfield Packing and walked into Johnson's office. She protested that she had gotten large raises every year and a glowing recommendation just two months before. "And now you're going to fire me? How many people have I fired?"[38] Buffkin has never clearly specified how many people she actually did fire, but there were so many that at one point company managers told her to stop firing employees for a while. To put this in context, as many as three hundred people were fired every month from the plant.

At this point, Buffkin and Johnson were locked in a brutal struggle, pitted against each other by the plant manager to whom they both officially reported. "Why are you firing me?" Buffkin said she demanded of Johnson.[39] "You fucked up this time. You messed with the wrong damn person." The reason for firing was left blank on Buffkin's termination notice. Buffkin has given a variety of theories for her firing. She has said that she believes Null told Johnson to fire her "because of her refusal to submit to Mr. Null's desire for a sexual relationship."[40] She has also said that she was fired because she had told them she wouldn't testify for them during the NLRB trial. "I told them I had lied for them for the last time, and I would not under any circumstances put my hand on the Bible and lie. At this point I was fired."[41]

Furious, Buffkin packed up all her belongings and left. She called Susie Jackson, crying and venting and threatening to testify against the company. She swung from anger to fear, telling Jackson she had a daughter to support. Buffkin was the big earner in the family; her husband, who worked in sanitation, made far less. She was afraid they would lose everything. In any case, if they had not fired her, she would not have decided to testify for the NLRB, she said. "I gave them fair warning. When I tell you something, if I ever see your fat ass, you better run. They did not believe I would do it."

One week later, Buffkin called a union attorney, who referred her to the NLRB. On October 8, 1998, lead NLRB prosecutor Jasper Brown went to Buffkin's house.[42] Getting a top-tier manager to testify would be a coup, and even though she had initiated the call, he wasn't sure she'd talk. Still, the tall, lanky attorney climbed into his old, comfortable Cadillac and headed deep into Bladen County, getting lost for a while down the unfamiliar winding roads. "I was skeptical at first," he recalled.

Brown, who retired in 2014, worked for the NLRB for forty years, most of them in NLRB Region 11, headquartered in Winston-Salem. Investigators from this office had worked the cases at the J. P. Stevens textile plants in the Carolinas in the early 1970s that inspired the 1979 film *Norma Rae*. In the movie, Sally Field starred as the young woman who holds up a piece of cardboard with the word "union" and starts a revolution. Sherri Buffkin and "Norma Rae" (in actuality Crystal Lee Sutton) could not have been more different in their attitudes toward organized labor. Buffkin didn't give two shakes about what she considered union nonsense, but her politics didn't stand in Brown's way. While Buffkin was a survivor and perhaps an opportunist who was cornered into doing the right thing, possibly for the wrong reasons, Brown didn't care, as long as she told the truth.

The first meeting with Buffkin lasted between two and three hours. On the one hand, she was guarded, just responding to questions. On the other hand, Brown said, she was forthcoming, especially about the firings of Ada Perry and Margo McMillan. Buffkin never told Brown about Null's harassment, but, Brown recalled, she was very worried that the NLRB and the union were in bed together, and she wanted nothing to do with the union. Brown explained that he didn't work for the union; he worked for the people. His goal was to make sure both the company and the union obeyed the law.

Buffkin agreed to testify. "I didn't have to do a big selling job," Brown said. "It was interesting because I was asking a supervisor—from a small town, white, rural, where everybody knows everybody—to go up against the big company. I think she was angry over being fired."[43] Buffkin said that she agreed to testify because she was tired of being a pawn. It wasn't for the union. It wasn't for the company. It was for the truth. "I was tired of all the lies."[44]

Buffkin gave Brown what he needed—a window into a world where supervisors were told to make union-supporting employees miserable so they would quit, or remove them from the general population to somewhere they could do less harm, or fire them. If they wanted overtime, supervisors were to deny it. If they didn't want it, they would have it piled on. If they had cleaned an area, make them clean it again.

As she spoke with Brown, Buffkin's feelings of guilt began to ease. "I wanted to be able to look my 10-year-old daughter in the eye with a clear conscience. Too many days I'd come home from work crying and my daughter would ask, 'Mommy, who did you have to fire today?' Smithfield Foods ordered me to fire employees who supported the union and told me it was either my job or theirs."[45]

Going into the trial, Sherri Buffkin was Brown's star witness, providing "key" testimony in the case.[46] But the prosecutors did not take her for granted. Two witnesses had already been "bought off," Brown said, and he very much needed a witness who could testify about the systematic illegal union-busting he knew was going on at the Tar Heel plant.[47] Buffkin gave him that and more.

## CHAPTER 9

# The Trial; A Surprise Witness

The trial was a whopper, even for a labor prosecutor with twenty-four years of experience like Jasper Brown, but the idealist—southern born and raised—was driven to right the wrongs done to workers. It was fueled by a coming of age during the civil rights movement. "I went to Booker T. Washington High School in Columbia, South Carolina," Brown recalled, in his gravelly low drawl. Each year, he and about five hundred other graduates return to the site where the two-story brick building was demolished when the school closed in 1974. All that remains is the auditorium, now owned by the University of South Carolina, where black students were not permitted to enroll until September of 1963. Once the state's largest all-black public high school, it has been called the "red brick oasis from discrimination" for Columbia's African American children.[1]

Brown went on to attend Hampton Institute, now known as Hampton University, in Virginia, a top-ranked Historically Black College and University (HBCU). A die-hard alumnus, Brown sets up his red canopy in the parking lot every homecoming and serves up fried fish and ribs with his son, who attended the University of North Carolina.

When Brown was first hired at the NLRB Baltimore office in 1974, he was one of the few African American prosecutors, as he had reminded them in very official ways, filing a federal Equal Employment Opportunity (EEOC) complaint in the mid-1970s over the lack of black NLRB attorneys in the region. On the surface, he seemed like an easy-going, good-natured, story-telling fellow, but in practice he was a maverick who was hard core when it came to the violation of workers' rights. Smithfield, he believed, was among the most pernicious violators he had ever seen, right up there with Tyson, he said.

The case alleged unfair labor practices in 1993, 1994, 1995, 1997, and 1998, including 286 allegations of threats and 25 allegations of illegal firing or threats of illegal firing, assault, arrest, threats of violence, threats of plant closure, threats of firing, threats of discipline, interrogation of employees about union sentiment, prohibiting union supporters from getting signatures on union cards, creating the impression of or using surveillance on union activities, informing workers that it was futile to select a union, verbal assaults to dissuade union support, intimidation and coercion, confiscation of union literature, threats to withhold a pay raise, harassment, prohibiting wearing union paraphernalia, threats of pay loss, promised benefits for not supporting the union, restraint, and more. Smithfield's alleged misconduct covered two union votes, spanned six years, and included the firing of a dozen workers for supporting the union, inciting a riot or near riot, and ordering the assault on a union worker by company men.

One reason the case was so large was that the NLRB had been too slow to react to initial union complaints after the 1994 election. Brown acknowledged that the NLRB should have moved forward with the complaints immediately after the 1994 vote instead of consolidating them. The problem was, he explained, that the violations were continuing at such a steady and rapid rate that the case snowballed. Prosecutors just kept adding complaints right up until it finally came before an NLRB administrative law judge in 1998. By then it had been seven years since the first workers' rights were violated, and the NLRB had yet to provide remedy to any of the workers. "We bear responsibility for the length of time that it took for the workers to get justice," Brown said. "We probably should have gone to trial in 1994. The problem was we had a legal obligation to investigate all outstanding charges and they kept violating the law, so we kept adding allegations to the complaint . . . but I would agree that we should have just taken it to court."

The mammoth size of the case caused all kinds of complications. In addition to reviewing details of each case file, Brown and his two assistants had to coordinate and transport more than 35 witnesses to take the stand over forty-six days spanning nine months between October 1998 and July 1999, then prepare to and cross-examine another 95 of the company's witnesses.[2] In addition, they had to provide proof of every allegation involving misconduct by 107 supervisors.[3] The level of detail was overwhelming, and tempers often flared.

"It was tense," recalled Don Gattalaro, the second NLRB trial attorney.[4] Unlike Brown, with his slow southern drawl, Gattalaro was a fast-talking New Yorker with quick gestures. He had attended the State University of New York at Rockport, then got his law degree at Wake Forest in North Carolina in 1985. A latecomer to the law business after a stint with the U.S. Department of Veterans Affairs in government service, Gattalaro was about 50 when the trial started. A third attorney, Tim Welch, also worked on the NLRB side.

In NLRB hearings, there are three sets of attorneys: the government's, the company's, and the union's. The union is called "the charging party." The company is called "the respondent." Each party usually has several attorneys. On some days, Smithfield had as many as five attorneys present, each probably charging $1,000 to $1,500 per hour. "I wouldn't be surprised if the trial alone cost Smithfield close to $1 million," Brown said. It was worth every penny in saved wages, estimated on a base rate at the Tar Heel plant in 2005 of $8.60, versus a base rate of $11.10 at Smithfield's eight unionized slaughterhouses.[5]

Presiding over the case was John H. West, a longtime NLRB administrative law judge, appointed in 1981. Born in Mt. Vernon, New York, he was a career federal government attorney who received his BA from Boston College in 1963 and his law degree from New York Law School in 1966. In 1975, he was made an administrative law judge.[6] Tall and solidly built, West was meticulous in his review of evidence, keeping copious notes throughout the months of the hearing, often recalling things the lawyers had forgotten. The judge actively questioned witnesses and was liberal in allowing evidence to be heard, despite the sometimes testy objections of all three parties.

Lead counsel for the union was Renee Bowser, the only other African American besides Jasper Brown among the lawyers in the courtroom, a fact they were both aware of and at times discussed. Bowser, who lived in Washington, DC, had majored in political science at Bryn Mawr, then graduated from North Carolina Central University School of Law in 1982 and earned a LLM (Master of Laws) from the University of Wisconsin Law School. When she graduated, she became an NLRB attorney, then went to work as assistant general counsel with the UFCW.

Lead counsel for Smithfield was none other than Bill Barrett, who had been the on-site labor attorney at the Tar Heel plant since before it opened in 1992. The labor world is a small one, and the attorneys had prior history. Gattalaro said that he and Barrett had argued a past case and that Barrett had told him, "You're nothing but a fucking government lawyer." "As you might guess," Gattalaro said, "that didn't engender any friendly personal feelings toward him." Barrett quickly gained a reputation for "arrogance," while union attorney Bowser, according to Gattalaro, was nicknamed "the diva."[7]

---

The trial started on October 19, 1998, at 12:30 p.m. at the Courthouse Annex building on Smith Street in Whiteville, North Carolina, just south of Bladenboro. By October 26, proceedings had been moved to the more suitable Bladen County Courthouse in Elizabethtown, where Buffkin had grown up. The boxy, two-story red-brick building built in 1963 still overlooks a large green lawn divided by a long concrete sidewalk and shaded by large trees. At the time that the trial was held, courthouse security was far more relaxed than it is today, and attorneys for the union, company, and NLRB could enter from side doors, making their way to the second-floor courtroom past gray-streaked Carrera marble walls. The courtroom was grand, with paneled walls, dentil molding, and four large portraits of former judges—all white men—two on either side behind the judge. A bank of thin, tall windows ran along the left wall. A chest-level wooden wall separated the judge from the tables where the attorneys sat, with the NLRB on the left. Judge West was one cool customer, according to Brown, almost always wearing a poker face.

The NLRB charged that dozens of employees had been illegally threatened and that eleven had been fired for union activity, including Lawanna Johnson, Keith Ludlum, George Simpson, Chris Council, Fred McDonald, Larry Jones, Patsy Lendon, Rayshawn Ward, Margo McMillan, and Ada Perry. The company's defense was obvious: every worker had been fired for cause—Johnson for excessive absenteeism; Ludlum for missing work for court; Simpson for not wearing safety equipment, leaving meat on the bone, and cursing at a supervisor; Council for safety issues, including horseplay in competing to open every hog coming down the line; McDonald for cutting down a hog, causing it to drop on the floor, and for "bursting hogs," or cutting into the guts, causing the line to be shut down for cleanup; Jones for wearing unauthorized union stickers and clothing; Lendon for using racial slurs and handing out union materials on company time; Ward for assault and starting a riot during the 1997 union election; McMillan for rudeness in the laundry; and Perry for threatening co-workers in the laundry.[8]

As with many complicated civil-style cases, the days were long, as often a half dozen attorneys slogged through the mud and weeds of tedium, arguing evidence on attendance policies and plant expansions, and calling multiple witnesses each of whom testified about the tiniest details of each alleged threat and firing. While theoretically the union and the NLRB were on the same side against the company, their agendas could diverge, though mostly they combined forces. Conflict among all three parties and obstruction from company attorneys began before a single witness appeared, as the company repeatedly failed to hand over required evidence including attendance records, recordings made of workers, and videos shown at mandated-attendance meetings.

Judge West frequently showed his impatience with the seemingly endless legal maneuverings. Bill Barrett's favorite phrase as he described NLRB and union requests for documents was "fishing expedition." Tempers were already hot early that second day.

> BARRETT: There are approximately 20,000 active and inactive personnel files of Smithfield Packing and a somewhat educated . . . guess would be that half or better of the former employees of this Company suffered from attendance problems. Therefore, the only way to comply . . . is to do one of two things. Dump 20,000 files into a warehouse and let Counsel for the General Counsel review them, or cripple the Company's operation by (having employees go) through 20,000 files for copies of attendance discipline.

> JUDGE WEST: The HR Department didn't keep termination records in its computer?

> BARRETT: There is a log that is coded in very basic terms. They want copies of disciplines, suspensions. There is no easy way short of pulling the files.

Apparently frustrated by legal wrangling, Judge West ended up calling the first witness himself, former director of human resources Sherman Gilliard, who was in the courtroom. The judge asked most of the questions, including how personnel records were kept and how they could be retrieved.

Even after witnesses began to be called by NLRB attorneys, who presented their case first, the early hours of court proceedings were often filled with the latest arguments over the company's unwillingness to supply documents. Attorneys moved paragraph by paragraph through what had been requested and not turned over. Each point was argued in immense detail.

"I'll just say, with all due respect, I am not going to comply with any order to let them go through all the personnel files from the entire history of this plant," company lawyer Barrett said. "That's not going to happen." Judge West called his bluff, telling Barrett that he could delay the case in order to provide the documents. "Something is going to have to be worked out," the judge ruled. "The matters sought clearly relate to matters under investigation here." Files for 16,000 employees were placed in a warehouse, organized by year. The NLRB and union attorneys would go through the boxes at night after court and on the weekends, accompanied by security.

---

As the days went by, dozens of witnesses described life inside the plant, from the wet kill floor, with stories of injuries and slipping in blood, to Livestock, with fights over pro- and anti-union stickers on hats. Union supporters told of exemplary commendations suddenly replaced by reports of poor performance. The pattern was described and redescribed in minute detail, day after day, by witness after witness.

On some days, as many as ten or twelve attorneys appeared in court together, and whether or not they cared for each other, nearly every day they all ate at the same restaurant a half block from the courthouse. Its name has been forgotten, though the quality of the food has not: "Their sandwiches were the best," Brown recalled.

Several witnesses were dramatically promising, and attorneys spent a great deal of time preparing for their testimony: Sherri Buffkin, Larry Johnson, Jere Null, and Joe Luter.

Meanwhile, what had been emotional events in real life—from firings that ruined lives to riots and arrests at union votes—were blandly rendered, interrupted by seemingly endless parrying by attorneys delving into the minutiae of every detail of every moment. Loud objections and flaring tempers punctuated the daily grind of repetitive, detailed questionings by the three teams. There was little drama in the retelling. Even testimony concerning the 1997 vote was punctuated by only a few moments of drama, as lead players Chad Young and Rayshawn Ward gave their version of events.

"Somebody grabbed me from behind . . . in a chokehold," UFCW campaign leader Chad Young recalled. "There were racial slurs. Nigger lovers go home. I heard a couple of screams. . . . It was a riot, because we're being kicked, we're being punched, we're being backed by the crowd, pushed completely out of that room. . . . I seen Jonny [sic] Rodriguez, a [UFCW] representative out of Dallas being led out in handcuffs. I seen Rayshawn Ward literally being dragged out of there in handcuffs with his eyes swollen shut."

Ward told how he was punched, kicked, and sprayed with mace. "Danny Priest told me I was under arrest and he kneed me in my back," he recalled. The mace was bad. "It was burning. It was like real burning."

Jasper Brown coordinated the case for the NLRB and chose to hold back Sherri Buffkin until the end, hoping to give a dramatic finale to the government's case. After the judge heard from the fired workers and illegally threatened union supporters, Brown believed Buffkin would tie it all together, showing that upper management orchestrated the illegal acts.

---

On November 4, 1998, Buffkin was ready. In preparation, she had stopped by Leinwand's dress shop on Broad Street in Elizabethtown, where she purchased an olive-drab pantsuit. That morning at home, she added a beige camisole and an expensive necklace, then drove the fifteen miles along Highway 41 to the courthouse. She entered the back door and settled into the dingy witness room to wait, uncomfortable and alone. She was glad Jasper Brown was going to be questioning her first. "He was the one person in all that mess that I trusted," she said in 2016. "He was neutral. He had nothing to do with the labor union."[9] Brown, of course, knew that company attorneys would be coming after Buffkin with everything they had. "They knew she was crucial. They were going to come after her strong. Sure enough they did."

Indeed, they appear to have started before she even got to the courthouse. The night before her testimony, her husband received an anonymous handwritten letter at work accusing his wife—in great detail—of sleeping with several people in the plant. Brown was convinced Smithfield was behind it when Buffkin handed him the letter on the morning of her testimony and seemed to waiver about testifying. "They were hoping [her husband would] go home and whip her butt and she wouldn't testify. So she was nervous, smoking, the whole business. I asked for a recess," Brown recalled.[10] He got her to calm down and promise that she would testify. Settling into an old metal chair, reading the latest John Sandford novel, she occasionally looked out the windows onto Poplar Street in downtown Elizabethtown. "I read to get my mind off everything else, so when I was asked questions, my mind would be free and I would tell the truth. It was me and that book. That was it."

If Smithfield managers did send the letter, they had guessed wrong again about how Buffkin would respond to pressure. "I hadn't completely made up my mind, but when they gave my husband a letter talking all this trash and Davie brought it straight to me and said 'I know better,' that was the turning point. That right there was the moment that sealed their fate, when they did that shit to my husband and he ain't had nothing to do with anything."[11] While she was waiting to testify, Buffkin asked to use the restroom. In the hallway, she ran into company attorney Bill Barrett, whom she called a blond-headed Ivy League type, wearing his usual expensive suit and know-it-all attitude. "I have nothing to say to you," Buffkin told him, thinking of the letter she held. Barrett smirked. "I looked at him and said, 'You don't talk to me without another lawyer present.' I had that shit in my hand and it was a done deal."[12]

Buffkin was summoned to the courtroom a little after the session began at 9:45 a.m. As she settled into the witness chair and looked out across a courtroom packed with union and company supporters, she suddenly realized there was no one there for her. She was not with the union. She was no longer with the company. She was all alone, right there in the little town where she'd grown up. "The union was there for the union. The company was there for the company," Buffkin said. "I told my family not to come, because I knew the company managers were going to lie, and I didn't want my family to hear the lies."

After reviewing her background with the company, Brown began to lead his star witness up to the evidence company attorneys feared most: that Buffkin had personally witnessed labor attorney William Barrett, general manager Larry Johnson, and vice president Jere Null breaking federal labor laws and that all three had directly ordered her to break those laws as well.

Allowing a government witness to accuse a company's lead attorney of illegal acts had required the highest level of approval from inside the NLRB. "We had to get clearance from Washington on that," NLRB prosecutor Gatta-laro recalled. "The labor board doesn't like to go after an attorney. But we needed that to show the animus and hostility from her supervisors and their attorneys."

Company attorneys argued vehemently that anything they had told managers — including Buffkin — was protected by attorney/client privilege and could not be introduced at trial. Judge West nixed this gambit, saying, "It's coming on the record one way or another." He said he would deliberate the matter of attorney/client privilege later.

The testimony led to the most stunning evidence of company wrong-doing: overt orders by Barrett, Null, and Johnson to fire union supporters.

BROWN: Do you recognize Mr. Barrett here today?

BUFFKIN: Mr. Barrett is sitting here drinking water.

KATZ: Objection — attorney/client privilege.

JUDGE WEST: Overruled.

BUFFKIN: Mr. Barrett asked me if Margo McMillan was one of mine, meaning one of my employees. I did tell him yes, she was. Mr. Barrett told me that he had just left a meeting, an anti-union meeting, in which her name came up repeatedly. He then looked me in the face and told me, "Fire the bitch. I'll beat anything she or they throw at me in court."

JUDGE WEST: Repeat that, please.

As the courtroom fell silent, Buffkin repeated what he had said. She told how Larry Johnson approved McMillan's firing and how Jere Null had vetoed Buffkin's plot to save McMillan's job. She told how John Hall had lascivi-ously sneered that McMillan would have done anything for him to keep her job. And, that Johnson had ordered her to cover their tracks by putting dis-ciplinary letters in the personnel files of all laundry employees. She testified that Null had told her to plot a setup to snare Ada Perry, that John Hall had executed it, and that Larry Johnson approved Perry's firing. With that, the NLRB had presented firsthand eye-witness evidence that Smithfield Packing's upper managers and its attorney, Bill Barrett, had plotted to violate federal labor laws by firing employees for union activities.

The cross examination was brutal. Before Buffkin was fired, she had met with Bill Barrett and other labor attorneys to prepare affidavits for the NLRB hearing. The affidavits clearly stated that Perry and McMillan were fired for cause and Buffkin had signed the affadavits, confirming by oath that they were true. So she was guilty of perjury either when she signed them or as she testified. Company attorneys hammered this point home. Either way, they argued, she was a liar and not a credible witness.

"I want to sleep at night," Buffkin told the judge in her own defense. "I'm not getting anything out of this except to ease my conscience. I know what happened, and it wasn't right." Joel Katz, one of the Smithfield attorneys, badgered Buffkin. "Notwithstanding the fact that on two different documents you swore—under penalty of perjury—that you were telling the truth, today is actually the day you're telling the truth because you couldn't sleep at night. Is that right?"

"That is true."

"It has absolutely nothing to do with the fact that you were fired from the company, does it?"

"No, sir."

More than a dozen times Katz pointed out that Buffkin had signed false statements, and more than two dozen times he noted that she had perjured herself.[13] Finally, Judge West interrupted Katz and told him to let Buffkin speak. She said that attorney Bill Barrett knew the affidavit was false because he had written it. "That's the statement you signed, isn't it?" asked Katz. Buffkin replied, "That's the statement that Bill made that I signed, yes, sir. Because I had a job. I had a family and I know that you don't go against Larry Johnson or what I've been instructed to do by them."

When court adjourned that day, NLRB prosecutors were "euphoric," Gattalaro said. "We thought we had won the case when she was done," he recalled in 2016. "She had struck a blow against the company's credibility and their attorney's credibility. That was pretty big."

---

But Smithfield attorneys weren't finished with Sherri Buffkin. When their case began, their witnesses painted Buffkin as a vindictive, conniving social climber, driven by money and power, vulnerable to bribes, and capable of plotting to take the jobs of her superiors. One of their main witnesses was Buffkin's longtime confidante and one-time best friend Susie Jackson, who had risen from flipping burgers to the head of sanitation at Smithfield supervising forty employees under Buffkin's tutelage. After Buffkin was fired, Jackson was promoted to purchasing agent with a small raise, taking over some of Buffkin's old responsibilities and reporting directly to Jere Null.[14]

On January 28, 1999, Jackson testified that before Buffkin was promoted she had sabotaged her old boss, the former purchasing agent, by hiding supplies so that workers would run out and she would be blamed. She had ordered Jackson to help. It worked, and the woman was fired; it was all a plot by Buffkin to get her boss's job, according to Jackson's testimony. Jackson also testified that Buffkin filed for workers' compensation for a fake backache. She said that Buffkin was a liar with a terrible temper who fired so many of her clerks that managers asked her to stop firing people.[15] Jackson had worked with Margo McMillan and Ada Perry in the laundry and both, according to her testimony, were fired for cause, not for their union support.

NLRB attorney Gattalaro tried to undermine Jackson's testimony.

GATTALARO: Why were you her friend if she had all of these bad qualities?

JACKSON: It was two different issues. I knew her for how she was and I accepted her for what it was.

GATTALARO: She made you do things that you didn't want to do, right?

JACKSON: That was on a work relationship.

GATTALARO: Oh, okay. And you just forgot all about that in your personal life.

JACKSON: No, I mean . . . (it) didn't affect me when I was not at work.

Buffkin never forgave the betrayal, right up until the time Jackson was dying of cancer and wanted to see her. She refused.

Smithfield had more witnesses lined up to testify against Buffkin. Plant manager Larry Johnson—the man Buffkin regarded as her main rival—testified on February 19. Johnson had supervised the slaughterhouse portion of the plant, while Buffkin ran the boxing, packaging, shipping, and cleaning of areas not directly related to the animals. Both reported directly to Null.

Johnson said he didn't trust Buffkin. The reasons were numerous. She had gone to the beauty parlor with her daughter when she was supposed to be at work. She had asked one of her employees for prescription-strength Motrin (a brand name for the analgesic ibuprofin). She said she got hurt at work and falsely filed for workers' compensation. She yelled, slammed doors, threw phones, and swore—a lot. She gave two cordless phones and a golf umbrella to her own employees instead of keeping them to raffle away at the annual employee picnic. "She admitted taking the prescription Motrin and I told her how disappointed I was," Johnson testified. "She admitted to the phones and umbrellas. She felt that they were low-ticket items and I felt that regardless of the value they were meant for Family Day and I suspended her."

"She had done a good job for the company as a purchasing agent, but I felt that I couldn't trust her," Johnson said. Though he had written her a letter of recommendation about six months before the firing, Johnson said it was to "fob her off on some other employer."[16]

Johnson got tripped up during cross-examination by NLRB attorney Donald Gattalaro. Was the letter of recommendation a lie? No. Was it true? No. Was her firing a group decision? No. Did he make the decision on his own? No.

The outcome of the case would be determined by whether Judge West believed Sherri Buffkin or Larry Johnson. The judge made it clear he wanted to hear every bit of evidence. He answered "overruled" eighty-seven times during Larry Johnson's testimony on February 19, nearly every time any of the three parties objected and tried to stop portions of Johnson's testimony.[17]

During the final hours of his testimony, Judge West took over questioning Johnson himself.

JUDGE WEST: And your knowledge regarding possible misuse of items is limited to two umbrellas and one telephone?

JOHNSON: Yes, sir.

JUDGE WEST: The one telephone went to an employee, not to Buffkin?

JOHNSON: Yes, sir.

JUDGE WEST: So, it's not a situation, in your opinion, where she took something for herself?

West asked Johnson about his friendship with Ada Perry and about the employee who ended up with the umbrella Buffkin gave her. He asked him about Rayshawn Ward and Johnson's role after Ward was handcuffed.

The judge could be sly in exploring the character of a witness. He would become inexplicably interested in one portion of the testimony. For instance, he seemed fascinated by the story of Johnson firing a supervisor who was breaking the rules by smoking with his employees in his office shortly after his father died. West asked if Johnson's father was still living. Johnson said that his father died when he was 14 and that it "tore his life up." West said that his own father had died in 1989 and he had been very upset. Before he fired the supervisor, West wanted to know, had Johnson considered that the supervisor might have been upset by the death of his father? In the retelling, Johnson explained that initially he didn't realize the supervisor was smoking, and he had fired an underling. The supervisor then held out his own cigarette and said that it wasn't fair to fire the underling and not fire him. So Johnson fired them both.

West was also very curious about Johnson's response to Margo McMillan's petition with signatures from at least two hundred co-workers vouching for her character. West wanted to know if Johnson had considered the weight of such a petition. Johnson said that he was more concerned with how McMillan had managed to get so many signatures when she was supposed to be working. West was also curious why Johnson so readily believed Jackson's story that Buffkin was getting her hair done during work hours, while Jackson never told him the far more serious infraction, that she and Sherri had hidden inventory to get the previous purchasing agent fired. Judge West may not have shown any emotion from the bench, but his active role in questioning gave a clear indication of what he considered key points.

---

On February 25, company attorneys put plant manager Jere Null on the stand to further undermine the credibility of the NLRB's strongest witness. Even if Null only had the one alleged consensual sexual encounter with Buffin, one has to wonder what was going through his head as he testified against her. Further, if he truly had fallen head-over-heels in love with her—as Buffkin said—and had offered to leave his wife and take Buffkin back to Smithfield and give her a life of unimaginable wealth, how must it have felt to betray her? Perhaps self-vindication played a part on both sides.

As NLRB attorneys expected, Null denied every complaint. If his under-lings had approved a firing, he trusted them and rubber-stamped it, he said. He never ordered supervisors to punish union supporters by assigning them harsher work assignments or ordered rules enforced only for union sup-porters. He never told Buffkin to fire Ada Perry. He never approved Margo McMillan's firing. He never told Buffkin that her department was a hotbed of union activity. He never used racial slurs. He only tried to help Buffkin become a better manager, to lead by example, to teach her to run her divi-sions in a more professional manner. He never discussed the union campaign with Buffkin.

Null said that he was consulted about Buffkin's firing. At first, he testified, he was against it. He, Larry Johnson, and a purchasing agent from headquar-ters in Smithfield met to discuss the accusations regarding Buffkin's giving away the phones and umbrellas and taking the prescription-strength Motrin prescribed for one of her employees. The group jointly decided to suspend her for one week, then changed their minds when they heard that Buffkin had loudly called the purchasing agent "a bitch" during lunch with friends at Bridgeman's Barbecue in Lumberton. Null said that he changed his mind and told Johnson to fire Buffkin.

There were obvious problems with this version of events. First, the thefts were of inexpensive items and Buffkin gave them away. Second, if Buffkin was so untrustworthy, why was she left in charge of a million-dollar pur-chasing budget, given repeated raises and written commendations? NLRB attorneys wanted to know. Null blamed it on Buffkin's people skills. "Do you normally promote people who are not performing well?" NLRB attorney Tim Welch asked.

After Null's testimony, company attorneys lined up so many witnesses to attack Buffkin's character that the judge stopped the proceeding. Barrett defended the strategy, saying the company had learned of her intent to tes-tify only a week before the hearing, and they had not had adequate time to prepare. Buffkin's testimony had damaged the credibility and integrity of the company, so the company should be entitled to undermine her character as well.

Company attorneys asked to recall Sherri Buffkin to resume their cross-examination from her first appearance. Judge West was not pleased by this deviation from normal procedure. Company attorneys argued they had new evidence suggesting she was not a truthful person. "Her character and her credibility is in almost every issue in this case," Barrett argued. The judge allowed Buffkin to be recalled but ruled that questioning would be limited to evidence of misconduct discovered after she had first appeared to testify. When government attorneys called Buffkin, she was furious and initially refused to come in, saying her daughter was sick with a high fever. The first round of testimony had been traumatic. By now, Buffkin wished she had never agreed to testify. She had no income: Smithfield had "blackballed her," she said, so she couldn't get work elsewhere, and she was having trouble making house payments on her husband's maintenance-worker pay.

When Buffkin finally appeared on March 3, she brought her own attorney, Mike McGinnis of Elizabethtown, to look out for her interest. The judge would not allow testimony regarding who was paying his fees. Company attorneys suggested it was the union. Company attorneys wanted to know why she agreed to return to the stand. Buffkin said that union attorneys told her the reason: "It was pretty much that I had been trashed and I was to come back and any questions that were asked I was to answer." Still sitting stiffly upright an hour after entering the courtroom, Buffkin had yet to answer any question of substance owing to attorney arguments. She was not in a good humor, nor was the judge.

When the first questions immediately went back to her taking prescription-strength Motrin, Judge West interrupted, saying that just because she took someone else's prescription drug did not mean she was not a credible witness. "We are focusing on the issue of the witness's veracity," West said, stating that a woman who had illegitimate children was not automatically a perjurer. At this point, the animosity building throughout the trial exploded in the courtroom. Union attorney Renee Bowser accused company lawyers of "trashing" Buffkin. Company attorneys retorted they were "getting the truth out." When Bowser came back with "They wouldn't even know what the truth is," Barrett leapt up. "Excuse me, say that louder so we can hear it for the record." Bowser didn't miss a beat, turning on company attorney Joel Katz: "Like you cursed me out in the hall, do you want to put that on the record?"

"I'm really at my wit's end for someone to say, F-U-C-K you, out in the hall. . . . I'm tired of this man," Bowser said. Katz retorted, "We're tired of each other." Katz defended his action: "They have accused us of lying from day one. They have impugned the integrity of myself, this company and it gets to be too much. She has no evidence that any of our testimony is false. That's going to be your province to decide who is telling the truth."[18]

Next, Buffkin—who had handled millions of dollars in purchasing—had to explain taking gifts from vendors. She testified she reported all gifts as required and that she wasn't the only one to accept gifts; Null and Johnson accepted more and better gifts than she did. She had turned down offers of trips to Cancun, ski weekends, and beach fishing trips, most offered by label-and-glue manufacturers. When asked about giving away a donated phone, Buffkin said that it was a "push button cheap phone you can get at Dollar General." The umbrellas she gave away she replaced in the inventory for the raffle giveaway.

As was typical, every vendor and gift was questioned in minute detail. "The extent they went after her in the trial was overkill," Brown recalled. "To use her best friend, someone who'd swapped off on babysitting, and have her testify against her showed the extent they were willing to go to undermine the union and anybody who supported the union." Buffkin was humiliated by all the accusations. Former co-workers, family, and friends were gossiping that she had stolen company property and demanded prescription drugs from her employees. Even her mother asked if these things were true. Buffkin said that she couldn't go to the grocery store without people staring at her. Given the small-town mentality, the facts were lost in misinterpretation, exaggeration, and outright lies with every retelling.

---

Perhaps the oddest move by the company team was putting Smithfield Foods chairman and CEO Joe Luter on the stand on February 24, 1999. He was, of course, by far the wealthiest and most famous person to testify, and the interplay between Luter and Judge West was among the most fascinating moments of the trial. West clearly wanted to know what Joe Luter knew, when he knew it, and whether he directly ordered illegal acts. He was also curious about Luter's opinion of the NLRB.

At the time, Luter was the sixth-highest-paid executive in Virginia, with a salary and bonus of $3.8 million.[19] The 1999 cover of the company's annual report to shareholders would say: "We are proud to report that Smithfield Foods is now the largest hog producer and processor in the world." Seven years earlier, Luter was listed as one of Virginia's one hundred richest people, with an estimated wealth of $81 million.[20] By comparison, when Sherri Buffkin was fired, she was making about $70,000 a year.

The CEO was confident on the stand, often offering his opinion of the legal nuances of the proceedings. At one point, he contradicted UFCW counsel Renee Bowser.

BOWSER: I object on the grounds of hearsay.

LUTER: It's not hearsay. This is direct . . .

JUDGE WEST: Mr. Luter.

LUTER: Sorry.

JUDGE WEST: You're a lawyer also?

LUTER: No, I'm not.

Jasper Brown, who handled the cross-examination, later wondered if the company attorneys thought Luter would wow the judge. "I think he felt like, here I am, the big big owner of a multimillion-dollar company and the judge is going to be impressed that I showed up to testify."[21]

Brown quizzed Luter on why he didn't personally investigate the allegations that company managers tried to start a riot after ballots were counted in 1997. Luter said that he rarely went to the plant or spoke directly to management there.

LUTER: Because I consider most NLRB complaints to be a matter of—
to be frivolous, outrageous, not truthful, and that it's just standard procedure that unions use to harass a company.

BROWN: Do you feel that other complaints that have been made against this company, specifically with regard to the EPA and waste water treatment, that those have been frivolous as well?

LUTER: No, I would not use the word frivolous. I do get agitated when you have environmental laws and one standard is applied to private industry and a total different standard is applied to municipalities or governments that are operating waste water treatment facilities . . .

BROWN: So the only frivolous complaints you feel that are being made are the ones that are made by the National Labor Relations Board?

LUTER: No, I think the courtrooms . . . are full of frivolous lawsuits.

BROWN: Nothing further.[22]

Brown had stirred the pot with his questions. Luter, whether purposely or not, appeared to have offended Judge West with his low opinion of the NLRB and their cases. The judge began to question Luter directly about past charges of unfair labor practices against Smithfield Foods. The CEO said that he didn't know whether there were any, but that probably there had been.

JUDGE WEST: To your knowledge, after all appeals were exhausted, has it ever been determined that Smithfield Foods, its subsidiaries or its affiliates, have violated the National Labor Relations Act?

LUTER: I don't know.

JUDGE WEST: In your mind, a violation of federal law is a pretty significant thing, isn't it?

LUTER: Absolutely.

As the judge tried to get a clear answer, Luter interrupted several times, saying there were so many complaints filed by federal agencies against Smithfield Foods that "we settle them all the time, it's easier to settle than to fight, particularly with the government who has unlimited resources."[23]

Judge West next returned to the night of the arrests after the second day of voting at the Tar Heel plant to determine whether Luter had any role in the planning.

JUDGE WEST: On August the 22nd, 1997, at your Tar Heel plant sheriff's deputies were on site. In advance of them being on site, were you notified that they were going to be on site?

LUTER: No.

Luter was excused. Judge West was obviously satisfied that Luter had not personally orchestrated the riot that night.

---

The last day of the hearing was July 19, 1999, just shy of seven years since the plant opened. While both the NLRB and UFCW were hopeful regarding the ultimate decision by the judge, the actual ending of the hearing was anticlimactic. Despite thousands of hours of work on the part of witnesses, attorneys, clerical staff, and the judge, there was still no hope of final resolution in sight. A few workers who had been fired for union activities went to work for the union, at least temporarily. Others, like Buffkin, despaired of finding work. Some moved on to other plants. Others simply disappeared, presumably continuing to work to provide their families with food, clothing, and shelter.

Hope was not a word that came up. The attorneys, who knew the system all too well, understood that, given the voluminous record, the judge could take months, even years, to render a decision. They knew Smithfield's record for being aggressively litigious. If the judgment was not in their favor, Smithfield would appeal, appeal, and appeal again to the highest court, potentially delaying justice for years.

Seventeen months later, attorneys were notified that a decision had been rendered. It was ten days before Christmas, in the year 2000, more than eight years after the plant had opened.

## CHAPTER 10

# The Judge Rules

Today, Judge West—who has retired to North Carolina—won't comment directly about the case. But his ruling, a 436-page document released on December 15, 2000, spoke volumes. In the end, West agreed with union attorney Bowser regarding the truthfulness of the company's witnesses. West discredited ninety-four of Smithfield's ninety-five witnesses. That means he did not find their testimony truthful or, in judge-speak, found them "not credible." They were either lying or had remarkably faulty memories, noted West. They included a district court judge, a district attorney, numerous high-level managers, and outside vendors.[1]

More important, West ruled that Smithfield had committed "egregious and pervasive" labor law violations during the two unionizing campaigns. He overturned the results and ordered a new election. He ordered eleven workers rehired and given back pay: Lawanna Johnson, Keith Ludlum, George Simpson, Chris Council, Fred McDonald, Larry Jones, Patsy Lendon, Rayshawn Ward, Margo McMillan, Tara Davis, and Ada Perry. He found that the company had illegally used a private police force to keep workers from unionizing, illegally fired workers, assaulted workers, incited a brawl after the 1997 union vote, and illegally threatened immigrant workers with deportation.

Smithfield created a "violent atmosphere" during the 1997 union vote count, West wrote, concluding that Rayshawn Ward had been assaulted and arrested because he was a union organizer. "Management planned, orchestrated, and incited the violence," he said, blaming Larry Johnson for egging on the confrontation and Jere Null for leading the melee. "If the testimony of the company's witnesses is to be believed, one has to accept as fact that Ward, who weighed about 110 pounds, for no apparent reason attacked [a deputy] who weighed 220 pounds, and [a co-worker] who weighed 270 pounds. Such a scenario, where Ward and his friends were overwhelmingly outnumbered, simply defies logic."[2]

West repeatedly credited Sherri Buffkin's testimony about the managers' anti-union acts. NLRB prosecutors were thrilled with the ruling and proud of Buffkin. "Jasper had prepared her well," Gattalaro said. "She never broke down. She was very good on the stand. We were worried because they were going to be attacking her personally."[3] Buffkin felt vindicated, and for the first time in two years was glad she had decided to testify. "The judge discredited every statement Jere Null made," Buffkin recalled. "That justified my life. No one knew what I had been through and how much my family has suffered. The only satisfaction came from Judge West. I read the entire four-hundred-plus-page decision and he never discredited anything I said. . . . No one can dispute what the judge said."[4]

---

Smithfield Foods, unsurprisingly, did dispute it and intended to fiercely fight Judge West's order to vacate the results of the two previous elections and his requirement that they hold a new, more carefully monitored vote. They appealed first to the full board of the NLRB. Meanwhile, the enforcement of West's order was stayed.

While 90 percent of NLRB cases at that time were settled once a decision had been rendered, Smithfield chose to fight while the workers continued to wait for justice. And justice from the NLRB is typically not swift. Buffkin, Ward, and Rodriguez all took matters into their own hands. Buffkin filed an EEOC sexual harassment suit against Smithfield Packing and Jere Null in 2001. Ward and Rodriguez filed a civil suit in federal court against Smithfield Packing and Danny Priest in August of 2000, claiming the company had violated their First, Fourth, and Fourteenth Amendment rights. They also brought state claims of false arrest, malicious prosecution, and assault and battery against the company and four deputies, including Danny Priest.[5]

Initially, Buffkin's suit was dismissed, but it was refiled in 2002. At a later, undetermined date, Smithfield Packing appears to have settled for tens of thousands of dollars. The suit alleged that Null kissed or attempted to kiss Buffkin in the hydraulic room, the conference room, his office, the engineering office, and the elevator. It alleged that Null repeatedly called or paged Buffkin, leaving coded messages. A copy of the suit was provided by the court and the NLRB, but it was also referenced in the two ongoing court proceedings: the appeal of the NLRB ruling by Smithfield Packing and the civil rights suit filed by Ward and Rodriguez. The appeal argued that the NLRB ruling should be overturned because Buffkin's EEOC suit proved that she had an ax to grind and thus her testimony was tainted. Attorneys in Rayshawn Ward's civil rights suit brought it up when they agreed that the sexual harassment lawsuit would not be mentioned during the trial.

The civil rights trial in the U.S. District Court for the Eastern District of North Carolina included many of the same witnesses who had appeared in the NLRB trial, and much of the testimony was repeated.[6] The initial result was much the same. In March 2002, a federal jury found in favor of Ward and Rodriguez and ordered Smithfield Packing and Danny Priest to pay $755,000 for civil rights violations. Ward was awarded $600,000 in compensatory and punitive damages. The estate of Rodriguez—who had died in December of 2001 — was awarded $155,000.

The response was jubilant. The UFCW put the word out in a news release:

> In a throwback to an era of hooded night-riders, brutal beatings and false arrests, a jury in federal district court in Raleigh, North Carolina last Friday found Smithfield Packing in violation of the federal civil rights law. . . . Smithfield had waged a vicious anti-worker campaign and created an atmosphere of racial hostility. . . . During the union drive . . . Smithfield held separate meetings for black and Latino workers to pit worker against worker based on race. On the day of the election, deputy sheriffs, dressed in battle gear, lined the long driveway leading to the Bladen County plant. The sheriff's menacing presence created a violent mood for the workers who were merely trying to exercise their right to vote for a voice on the job.[7]

The money was never paid. On July 30, 2003, a federal appeals court in Raleigh overturned the jury's finding because on the night they were arrested the two men had signed waivers of liability to avoid criminal charges and had promised not to sue deputies for their arrests or injuries. The logic was based on what Danny Priest was wearing the night of the union vote: his Bladen County deputy's uniform and badge. The appeals court determined that Priest—though head of security for Smithfield and head of its private police force—had made the arrests as a deputy; therefore, he was covered under the waiter the two men had signed. Smithfield was released from liability because, the appeals court ruled, the company did not control or set policy on state arrests, despite evidence that deputies took the men to Jere Null's office first rather than directly to jail; and testimony during the NLRB hearing showed that the officers spoke with Null for about thirty minutes before formally charging Ward and Rodriguez. Consequently, Judge West concluded that Null controlled the deputy's actions. But the appeals court did not agree.[8]

---

While both civil cases were moving through the courts, Smithfield Packing was arguing their appeal to the full NLRB board. They filed their seventy-three-page formal exception to Judge West's decision on April 6, 2001, which included 490 individual exceptions to the judge's rulings, including whom he found credible and the remedies he had recommended to ensure a new election would be fair and legal.[9]

Smithfield's main argument was that Judge West "relied on only one witness, Sherri Buffkin . . . a former supervisor with a severe bias against the Company,"[10] while crediting all the government's witnesses and discrediting all of the company's.[11] "This remarkable set of credibility findings, standing alone, casts serious doubt on the reliability of ALJ [Administrative Law Judge] West's analysis in its entirety," attorney Willis H. Goldsmith wrote to NLRB counsel.

Buffkin was lying, according to the 124-page brief in support of the exceptions, because she had filed a "bitter lawsuit" claiming she was fired for refusing Jere Null's alleged advances, not because she wouldn't testify for the company during the NLRB hearing. No evidence of that had been offered other than her own testimony that she just "knew" not to go against management, the brief said.[12] They argued that Buffkin's testimony should be thrown out.

The brief laid out arguments, debated at length during the trial, to the effect that all eleven employees were fired for cause, not for union activities. Five were discharged for "egregious violations of the Company's attendance policy and the other six were discharged for other misconduct." Attorneys argued that the judge was "apparently irked by [the company's] decision not to produce the videotapes [of Null's twenty-fifth-hour speeches]," and that led him to discredit Null's testimony about what he said during those speeches. Finally, attorneys argued that the judge's remedies were "extraordinary."[13]

And indeed they were. The judge had ordered the new election to be held outside the plant, on neutral territory. He had ordered that a notice be published in newspapers, sent to all who were employed there since 1993, and posted in Spanish and English inside the plant. He had ordered Null to sign and read the notice aloud, or be present while a board agent read it. In addition, he had ordered the company to turn over the names and addresses of all current employees and give professional union reps access to nonworking areas of the plant, including cafeterias, rest areas, and parking lots. Further, he had ordered that any time the company discussed union organizing, union reps were to be present and given equal time to respond.

Ironically, attorneys argued that the judge's decision should be overturned because the company had "*never* been found guilty of an unfair labor practice at the Tar Heel facility" and Smithfield Food's subsidiaries had been found to have violated the NLRA only three times, all three violations minor and quickly resolved. Finally, they argued that the guilty acts of a few in the context of thousands of workers and supervisors—including turnover of more than 16,000 employees in the six years preceding the trial—constituted a tiny percentage of conduct overall. The obvious flaw in the logic, of course, is that the eleven firings were not the only illegalities investigated, but only the ones the NLRB brought to trial, the ones where witnesses could be found and convinced to testify.

The company also filed motions, on April 5, 2001, to reopen the record on account of Buffkin's allegations that she was fired for "her refusal to submit to defendant Jere Null's sexual demands."[14] Documents, which may have included Buffkin's and Null's depositions in that case, were sealed by the district court on July 12, 2000.[15]

On April 24, 2001, NLRB attorneys responded to the NLRB, arguing that the newly discovered evidence in the EEOC lawsuit simply provided more detail about the underlying reasons for Buffkin's firing but did not alter those reasons substantially. "Obviously, Larry Johnson would not have informed her that she was terminated because she refused to submit to the sexual demands of Jere Null," they argued in the seven-page motion against reopening the record. NLRB attorneys also sought to downplay Buffkin's importance to the judge's decision, saying there were a lot of witnesses.[16] But none had been a high-ranking manager, none had exposed the motives of managers who willfully used the weak enforcement of the NLRB to their own advantage, who brazenly fired anyone who supported the union they were so determined to keep out.

———————————

While the NLRB appeals process played out ever so slowly—promising to take years—the union used Sherri Buffkin as their star witness in the public eye. On June 20, 2002, Buffkin testified before the U.S. Senate Committee on Health, Education, Labor, and Pensions hearing on "Workers' Freedom of Association: Obstacles to Forming Unions," chaired by Senator Edward Kennedy (D–Massachusetts). The committee was investigating and debating increased corporate power in the workplace—including the use of union-busting consultants—and the decreased worker's right to collective bargaining through a union. It was the first committee in fourteen years to explore possible solutions to dysfunctional labor law.

Buffkin told the committee that she was forced to get rid of union supporters. "Smithfield Foods ordered me to fire employees who supported the union and told me it was either my job or theirs." She testified that the company had separated workers by race and pitted them against each other. "The word was that black workers were going to be replaced with Latino workers because blacks were more favorable toward unions." Buffkin said she came to realize that Smithfield would stop at nothing to keep the union out. "I'm here to spread the truth and ask you to go see for yourself. The workers at Smithfield need your help," Buffkin told Senator John Edwards (D–NC).[17] The late Senator Paul Wellstone (D–Minnesota) and Senator Tom Harkin (D–Iowa) publicly thanked Buffkin. Senator Harkin called her testimony "a real profile in courage."

"You have just showed tremendous courage in what you have done," Wellstone said. "I think you will light a candle for a lot of other people and I would like to thank you." Wellstone spoke directly to Buffkin's daughter, Nicole, several times, telling her that her mother was special and that everyone was very proud of her. Buffkin still has a letter from Wellstone in her dresser drawer. "Thank you for your courage. I will not forget your testimony. I will work for the challenge. My best to you, Paul." Kennedy thanked Buffkin for exposing "one of the nation's most disgraceful and undemocratic situations, a company-sponsored union-busting campaign that destroyed the lives of hundreds of hard-working people." He told her daughter Nicole that her mother was a hero.[18]

But Congress took no action; the workers at Smithfield Packing continued to wait for justice, which came in fits and starts, incrementally, through the broken system of the NLRB.

---

On December 16, 2004, four years and one day since Judge John West had rendered his decision condemning the actions of Smithfield Packing managers, NLRB members Robert J. Battista, Wilma Liebman, and Dennis P. Walsh adopted nearly all his findings, including that Smithfield illegally tainted election results by intimidating union supporters through firings, surveillance, arrests, violence, and threats of violence. The board refused to reopen the record to admit additional evidence about Buffkin's sexual harassment lawsuit: "We find that the evidence . . . would not affect the judge's finding that Buffkin was a credible witness at this hearing, and we deny the [company's] motion." They also refused to reverse the findings on the basis of Smithfield's charge that Judge West was biased. The board said that the allegations were "without merit."[19]

It had taken them nearly three years and eight months to consider the arguments, review Judge West's decision, and render a final decision. It took Smithfield less than two months to file its appeal to the federal circuit court in January 2005. The clock reset, again.

It was very rare at that time for companies to appeal, which indicates how determined Smithfield was to delay justice and keep the union at bay. At the time, only 10,000 of the 30,000 complaints filed with the NLRB each year, or one-third, were found to have merit. Of those, 90 percent—about nine thousand—were settled. Smithfield was one of only about one thousand cases—or .03 percent of cases filed—to be appealed.[20]

Smithfield officials said that it was a matter of principle. The NLRB was dead wrong. "The company looked into the allegations and determined they did not engage in the conduct the [NLRB] judge says they did," said Smithfield's lawyer, Greg Robertson, who replaced William Barrett.[21] Tar Heel workers didn't want a union, said Robertson; if they did, they would have voted yes during union votes there in 1994 and 1997. Plus, times had changed.

"You cannot draw extraordinarily sweeping conclusions from events in 1994 and 1997 and say it continues today," Robertson said. "It's eight years later. It's a different place with different people and different employees."[22]

Yet some employees remained, and it appeared to be business as usual. Danny Priest had been made chief of the special police, despite conclusions by the NLRB that he illegally used the company police force to intimidate workers.[23] General Manager Jere Null and Plant Manager Larry Johnson had been promoted to corporate vice presidents, despite Judge West's determination that both had lied during the trial and that both had ordered union workers fired and incited a riot that resulted in the assault and arrests of two union supporters.[24] Robertson swept all that under the rug. "In the end, each party has to review the facts and look their own people in the eye and make the decision they believe is right," Robertson said. "We've tried to learn from what people think was going on, make judgments and decisions from there. To make it seem as if it were all still happening today is not the company's view of things. This all deals with what happened in the past."[25]

Union attorneys were pessimistic about delayed justice from the NLRB. "Those workers won't want to come back," said Renee Bowser, the UFCW attorney; "Smithfield doesn't lose." The NLRB—despite its exhaustive investigations and scathing conclusions—had thus far proven itself powerless to stop the company's exploitation of workers some twelve years after the first workers were fired. "We think it's an outrageous amount of time," UFCW official Tom Clarke said in 2005.[26]

During that outrageous amount of time, Smithfield had been taking over the industry.

The purchase of John Morell & Co. in 1995 had propelled Smithfield into its position as the leading pork processor in the United States. By 2000, it was also the world's largest hog producer, after purchasing Circle Four Farms in 1998, fourth largest hog producer Carroll's Foods in 1999, and the country's largest hog producer Murphy Farms in 2000. The acquisition nearly doubled Smithfield's sows (from 360,000 to 685,000) and gave it 60 percent vertical integration, *The Daily Press* reported, "market control from the squeal to the meal."[27]

In 2001, Moyer Packing and Packland Holdings became Smithfield Beef Group—the fifth-largest beef processor in the country. Farmland Foods was purchased in 2003, and two English companies were purchased the following year.

By 2004, the company controlled 27 percent of the U.S. market share, raising 12 million hogs and processing 27 million a year. Smithfield stockholders were earning 43 percent returns, according to *Fortune* magazine's April 18 edition, which ranked Smithfield second out of the nation's top twenty food consumer products companies, after Hershey Foods. Between 2000 and 2005, revenues doubled from $5 to $10 billion.[28]

In June 2005, the company announced record sales of $11.4 billion for the fiscal year, up nearly 23 percent from 2004 and more than double 2000. Operating profits were up 58 percent for the year to $582 million. Net income went from $227.1 million in 2004 to $296.2 in 2005.[29]

At the Tar Heel plant alone, the company was saving an estimated $34 million every year in wages and benefits, based on six thousand workers at base rates of $8.60 an hour versus $11.10 at Smithfield's other UFCW plants. In the end, as critics of the NLRB had pointed out, labor fines and penalties were puny, a paltry cost for a business fat with profits.

But the union said the company's fight was not just about profits. It was about power.

"They want ultimate control and are willing to enforce it," said UFCW official Jim Papian, echoing Smithfield plant managers who had threatened to throw workers back into the cotton fields they had come from. "The way Smithfield treats workers with their plantation approach shows basic disrespect for the people who work there and for the larger community."[30]

## CHAPTER 11

# On the Road with Union Organizers

B y 2005, unions in the United States were in major conflict, unable to agree on priorities or strategies. As the American Federation of Labor and Congress of Industrial Organizations (AFL-CIO), the largest federation of unions in the United States, prepared for its fiftieth anniversary in Chicago in late July, several of its biggest members—including the UFCW—split away to create a new federation, called Change to Win. Shortly afterward, other unions followed.

The split was over growing differences in strategies for the future. Those who supported Change to Win believed unions needed to focus their money on organizing workers. They believed the AFL-CIO had lost its way, spending too much money on national elections, specifically in supporting Democratic candidates, such as John Kerry in 2004, when a large minority of union workers were voting Republican. Change to Win was founded on the idea that the unionization of millions of workers had to be the number one priority, requiring that at least 75 percent of its budget be funneled directly into organizing.

AFL-CIO president John Sweeney (president from 1995 to 2009), had swept into power by promising to use, he said, "old-fashioned mass demonstrations, as well as sophisticated corporate campaigns to make workers' rights the civil rights issues of the 1990s."[1] Sweeney had made his name with aggressive tactics even before becoming president of the national Service Employees International Union (SEIU), when he preached aggressive organizing and confronting corporate power while building coalitions with other like-minded organizations. His successful SEIU campaign, Justice for Janitors, came in the late 1980s. His biggest success in his new role with the AFL-CIO was the Teamsters' nationwide UPS strike in the summer of 1997.[2]

Sweeney wanted to put intense economic pressure on entire industries, strong-arming them into accepting unions across the board, rather than using the traditional method of organizing one location at a time. He initially pressed for innovative techniques of using publicity to influence consumers and stockholders in corporate campaigns. But the AFL-CIO under Sweeney, according to Change to Win, had become politicized and was not focused on winning new union members.

Union members of Change to Win were called upon to escalate their organizing campaign. The UFCW's largest campaign was at Tar Heel, and it became the new federation's most important campaign. It was a call to action, but the challenges were great. The annual 100 percent turnover made it difficult to keep the union cards signed by workers current. The years of NLRB deliberations had left workers—and union reps—disheartened. And the company's shift from hiring African American locals to hiring newly arrived immigrants, many undocumented, meant the union itself had to unify its black and Latino organizers under one roof.

---

A new local campaign manager, Eduardo Peña, had been in place since the year before, when the UFCW was still part of the AFL-CIO, trying to reconnoiter ground troops and rebuild energy sapped by years of litigation. Peña, an organizer from Tampa born in Puerto Rico, had been working to rally plant workers who felt abandoned after the 1997 vote. Building trust and community—including overcoming deep-seated racism—within and around the plant was essential. But building solidarity between blacks and Latinos who lived and worked separately and didn't trust each other was difficult.

Peña moved the local headquarters from its old trailer in downtown Tar Heel into a new, one-story office in Red Springs, about a half hour's drive from the plant, to get some distance from surveillance and harassment by Smithfield Packing police.[3] The small gray office on Highway 211, with its sandy parking area, was easy to miss, but word spread among the Hispanic population that help was both available and free, and immigrant workers figured out how to get there—by the hundreds. Inside, colorful posters lined the walls: "Eastern North Carolina Workers Center"; "Centro de Trabajadores del Este Norte Carolina." UFCW newsletters were stacked on shelves: "Join the Campaign: Justice for Smithfield Workers"; "UFCW: A Voice for Working America."

The new office—instead of only being about unionization—was now a workers' center, where organizers guided immigrants through the maze of paperwork, including how to claim benefits and workers' compensation and disability, and helped them file lawsuits. Worker injury was quickly becoming the most important reason workers needed a union, Peña said.[4] His team ranged between six and fifteen people, a revolving group of former workers, experienced organizers, and new volunteers. It was David and Goliath, with an inexperienced crew up against a sophisticated company that had been running an aggressive campaign against the union for years.

In mid-February of 2005, Peña invited Human Rights Watch (HRW) investigator and adjunct labor professor at Cornell University Lance Compa to speak to workers at the local NAACP (National Association for the Advancement of Colored People) headquarters in Fayetteville. A few weeks earlier, HRW had published Compa's report *Blood, Sweat and Fear: Workers' Rights in U.S. Meat and Poultry Plants*, which investigated practices at Nebraska Beef, Tyson Foods, and Smithfield Foods' Tar Heel plant in 2003 and 2004. Among the three plants, in Compa's opinion, Smithfield stood out. "Smithfield is the central case study of the report because all the issues we looked at —worker freedom of association, health and safety, the right to workers' compensation, the exploitation of immigrant workers, generalized fear of the workers—all these are present. Also . . . the sheer size and scale of the facility and its importance in this industry made it a focus."[5]

The report focused on health, safety, workers' compensation, organizing rights, and rights of immigrant workers. Compa told the assembled group that workers were frequently injured, then refused medical care and workers' compensation, or fired. Injuries at the plant were common, with repetitive-motion injuries most frequent and knife wounds coming in second. All were underreported by management. "Workers make—not hundreds—but thousands of cuts, using the same muscles," said Compa.[6] The average worker has just a few seconds to make precision cuts in carcasses flying by.

Compa had included much of the testimony from the NLRB trial in his report, but focused primarily on the high level of danger and risk of physical injury. "Management knows how to make the plant safer by slowing down the pace and giving more frequent breaks," he said, but they keep the lines running fast to maximize profits. Then, when workers are injured because of high line speeds, they are afraid to report injuries. Compa had investigated incentive bonuses given to managers whose workers had low injury reports.

Compa's audience at the NAACP that night was a mix of ages and races, including very young Latina girls, older Latina women, burly middle-aged black men, and older black women. Peña translated. While posters of Martin Luther King Jr. and other black leaders looked down from cracked plaster walls at the NAACP office, workers debated how to handle injuries. "If we report an injury, we'll be fired," one worker said. "If we don't, there's no record." Without a record, any workers' compensation claims will be denied, organizers explained. "File a report," one said. "If you're fired, we'll help you."

Most injured workers, especially undocumented immigrants, simply move on. But for those who seek compensation, Smithfield settles about 90 percent of cases, even for undocumented workers.[7] "I really can't understand why it is Dr. Jekyll at other Smithfield plants and Mr. Hyde here at Tar Heel," Compa said in an interview at a fast food restaurant after the speech.[8] Indeed, this was one of the lingering questions from the NLRB trial. Some believe it had to do with Luter's nature: he didn't like to lose, and the stubborn entrepreneur and aggressive negotiator simply would not back down. "Most people are unaware of what it takes to put meat on the table in industrialized plants," Compa said, pushing back his round metal-framed glasses and smoothing his hair. "The pace, the punishing physical labor, repeated heavy lifting, slippery conditions, repeated cuts, the hard-cutting motions that create repetitive stress injuries."

The morning after the meeting, nearly a dozen organizers, mostly Latino, met at the new union headquarters. They laughed and teased each other, aiming a few smart remarks at Peña, who had gotten some attention from the younger immigrants the night before because of his tall good looks and shiny black hair. "Ooooh," teased one middle-aged Latina organizer. "Those young girls, they looked at you like you were a drink of cool water on a hot day." Peña was also teased about the special espresso machine he kept in his room at the Holiday Inn in nearby Lumberton. He traveled with it, working all week, then headed home to his family every weekend. He probably needed the coffee: he was constantly thinking and speaking in two languages.

Peña listened as organizers worried about workers who had attended the Compa speech the night before. One worker had been quoted by a reporter. "Monitor what happens," Peña said. "Give him a call today. If his supervisors mention it, let us know." Also weighing on Peña was a mother with a newborn. "I met with a lady who had a baby March 7 of last year and took several weeks off, then got pregnant again right away," Peña said. "She went home from work last Friday, gave birth on Monday, and they told her to be back next week or she'll be fired. It's insane." Peña said that union reps would call the company, mention federal family leave law, and try to get her job back.

After reviewing union sign-up cards and reporting the previous day's contacts, union organizers headed into the community, often to make cold calls at stores, laundromats, supermarkets, trailer parks, and various Latino hangouts. They stopped their friends and people they knew. The idea was make connections, try to help. Organizers made home visits nearly every day, driving through the flat, isolated counties surrounding the plant: Bladen, Robeson, Cumberland, Columbus. They kept track of contacts on cards and rated them from one to four, ranking the level of union support.

"The company police don't intimidate them at home," said organizer Kevin Blair. The 27-year-old, who is white, graduated from high school in Fayetteville, left as soon as he could, and never thought he'd be back. While working at a restaurant, Blair joined workers in a collective action and said he found his calling. "There's nothing like the feeling you get when workers stand up for themselves," said Blair, a graduate of North Carolina State. "It's a fight. I'm in the fight."

Blair stopped in Fayetteville to see Ray Hall, a 42-year-old black man. Hall had sued Smithfield after he was fired from his job hooking hogs because he ruptured a disc in his back when he tried to catch a falling side of pork. Hooking hogs requires sinking a hook into a fifty-pound side of hog and sliding it onto an adjacent table, an action repeated about nine thousand times per shift. Hall had been making $11.60 an hour, plus health insurance, retirement, and discounts on meat.

According to Hall's medical documents, Smithfield said that the injury was not work related. Hall lost his job and his health insurance. When the company eventually settled out of court for $15,000, Hall said, most of the money went to pay for doctor's bills and a lawyer to get him on permanent government disability.[9] Hall believes he was targeted because of his race. "If you notice, the ones they're getting rid of are black. They'll get rid of blacks so they can hire Mexicans. They can do the same thing that will get a black fired and the Mexican will still be working. Mexican rules out there."

Even union organizers could be suspicious of each other along race lines. Organizer and former plant sanitation worker Dan English, who was 58 in 2005, is white and grew up in Robeson County. After a military stint, he went to work in Tar Heel. Because he supported the union, English had a "target on his back" after his picture appeared in a local news story when he attended a union rally in December of 2003. That same month, a manager told English he "would not touch him with a ten-foot pole," according to the NLRB, which concluded Smithfield illegally fired English because of his union activities.[10]

English, a gray-haired fellow with thick, working-man's hands, organized workers for the union, tooling around rural North Carolina in his small, dilapidated truck talking about the power of collective bargaining. "Sometimes I feel like Moses, crying, 'Let my people go!'" he said.[11] He drove through black and Latino working-class neighborhoods, noting the sharp contrast between a worker's trailer and plant manager Larry Johnson's lavish new brick home on Cromartie Drive with its custom pool, reminiscent, he said, of "plantation houses and slave cabins." The Cromartie family had been one of the larger slaveholders in Bladen County before Emancipation.

A bumpy dirt road led to a trailer park where Smithfield Hispanic workers lived. English compared the ubiquitous grills and satellite dishes of the new immigrants to the lifestyle of Smithfield's chairman and CEO Joe Luter III. "Luter makes $850,000 a year with a $6.6 million bonus. It's discouraging. It makes people feel beaten down," said English in his gravelly voice. "The company owns this county. . . . They'll do anything they can to keep the union from coming in."

English pointed to an old farmhouse across from the plant. "That's the McNair house. That's where they do the hiring. Now they're hiring nothing but Hispanics. There were no Latinos here when I grew up. . . . The schools were segregated." He said that he understood the racial uneasiness at the plant, pointing to the rapid-fire Spanish around the organizers' table that very morning. "I know them and trust them, but it's still hard not knowing what other people are saying, maybe about you," he said.

Indeed, organizers were finding it nearly impossible to unify whites, Latinos, and black workers. "Long before the Mexicans arrived, Robeson County, one of the poorest in North Carolina, was an uneasy racial mix," wrote *New York Times* reporter Charlie LeDuff in 2002, after working at the plant. "Until a dozen years ago the county schools were de facto segregated, and no person of color held any meaningful county job from sheriff to court clerk to judge."[12]

St. Pauls—about fourteen miles from Tar Heel—was a watering hole during the 1700s along the stage road that ran between the Great and Little Marsh Swamps halfway between Fayetteville and Lumberton, where stage-coach drivers changed horses at the livery stable. The Presbyterian Church was built in 1799, and the mainly Scottish settlers built around it, with names like McNair, McNeill, McEachern, McArthur, McLean, McGoogan, and McKinnon.[13] In 1908, the first mill, St. Pauls Cotton Mill Co., opened. By 1920, two more mills had been built, along with a flour mill, a bank, a news-paper, an ice company, a lumber company, and three garages. In 1943, Bur-lington Industries bought the local mills and enlarged them, but business soured leading into the next century. In 1997, Carolina Mills bought the plants from Burlington, then closed its yarn-spinning plant in 2000, owing to a decline in orders and foreign competition, which cost two hundred jobs.[14] By 2001, Carolina Mills closed its other plant, eliminating another 320 jobs. The Cape Fear region lost nearly 10,000 textile jobs in five years.[15]

Residents deserted the old mill houses. Restaurants emptied out. Store owners struggled. "But just as the town seemed poised to plummet toward a less-than-pleasant fate, something stopped the free-fall," the local paper reported. "Immigrants from Latin America, lured north by the promise of jobs, began arriving."[16] Even long-settled families could not ignore the buying power of immigrant meatpacking workers, who were reviving the economy. A new town seal touting "The Little Town with the Big Heart" replaced the old one that said "The Textile Center of Robeson County." It promised welcome to the waves of arriving immigrant workers.

The first to arrive were the single young men. Then families came and stayed. They bought homes and sent their children to school. Between 2000 and 2010, the population of Hispanics went from 306 residents to 511, a quarter of the town's population of two thousand. "The Hispanics really stepped up when the mills left," said Paul Terry, editor of the *St. Pauls Review* and president of the Chamber of Commerce. "They opened mom-and-pop stores and they kept our mom-and-pop stores solvent. . . . In an effort to find a better life for their families, they helped make life better for an entire town."[17]

But not everyone was happy. Retired men who met to eat breakfast at Hardee's talked of illegals, increased drug trafficking, unemployment, and stolen jobs. They expressed anger about unlicensed and uninsured drivers and having to hire interpreters for the police department. "There have been hiccups in the town's transition from insular textile village to mini melting pot."[18] Bridging the gap between the races throughout the community was key, especially in the plant. But union organizers were losing the ground game. The 100 percent annual turnover of workers and the constantly arriving immigrants — many undocumented and wanting to keep a low profile — made keeping union sign-up cards up-to-date nearly impossible.

Workers on the ground were tired and frustrated with the pace and scope of the union's movement to unionize Smithfield Packing. It was time for the UFCW and the Change to Win federation to make good on its promises to hold global corporations like Smithfield Foods accountable.[19]

## CHAPTER 12

# The Corporate Campaign;
# Basic Human Rights

Though the UFCW had won two major victories with the NLRB—first in court with Judge West in 2000 and then on Smithfield's appeal to the NLRB board in 2004—the giant pork company, with its expensive team of lawyers, was successfully delaying final implementation of the judge's orders through the U.S. court of appeals. Winning battles in court was clearly not enough and was taking far too long. High-ranking union officials knew they had to take their case to the court of public opinion, and the best way to do that was to wage a national "corporate campaign" that would bring negative attention to the company's illegal practices. But a large-scale corporate campaign is not easy to run. The UFCW had to find just the right person, someone with experience who was intelligent enough to mastermind a multipronged and carefully timed attack. Someone who could size up the strengths and weaknesses of the global conglomerate and estimate—then counter—its responses. Someone with the charisma to partner with and harness the power of organizations outside the pork industry. Someone savvy enough to manage a public relations campaign that would put enough economic pressure on the massive company to make it see the light.

That's where Gene Bruskin came in. Bruskin, then 60, a career union organizer with eleven years at the AFL-CIO, had been secretary-treasurer for the Food and Allied Service Trades Department of the AFL-CIO since 1994 and labor director of Reverend Jesse Jackson's National Rainbow Coalition for two years before that. But he had more than experience. He had a fierce passion fueled by his belief that the common worker had the right to decent working conditions, decent pay, and the right to join forces with fellow workers to level the playing field with corporate fat cats.

Gene Bruskin at his home in Maryland in 2016, holding up just part of the documentation he has saved. Photo by author.

Bruskin had been raised in the steam of the 1960s, fed on the fever of antiwar activism, inspired by solidarity with the Cuban revolutionary Venceremos Brigade, and fueled by adrenalin on the picket line as a young school bus driver. His northern city blood ran thick with union revolution, inherited from his radical communist father. The Tar Heel campaign was to become his obsession, a challenge taken on during the peak of his long career. Hired with extra money as part of the call to action from Change to Win, Bruskin brought with him a belief that union campaigns were won from the ground up, getting buy-in from workers and community, church, and political leaders. When he drove to the new union office in Red Springs that January of 2006 in his new Ford Fusion, Bruskin was in top form and ready for the long fight.

As campaign director, he was about to take over one of the largest, longest, and most contentious labor campaigns in the history of the UFCW, as well as one of the most important union campaigns of the latter half of the twentieth century. Bruskin would have help from his teams in Washington and North

Carolina and his UFCW bosses, but the antiwar activist—and former United Steelworker—was the field general. And he was about to change the rules of engagement into fierce hand-to-hand combat against a corporation whose moral compass—and fighting style—was set by Joe Luter, the man on top at Smithfield.

Despite their differences, the two men shared some commonalities. Like Luter, Bruskin grew up working class but in an urban setting in the North, not a small town in the South. Bruskin's family, who had settled in South Philly, weren't religious, but they strongly identified culturally as Jews. "My great grandparents in Russia had a hat business for the Cossacks. One day a pogrom threw them down a well. My grandfather escaped and opened a hardware store in Philadelphia."[1] The hardware store is still there a hundred years later, and family members still live above it, just as Bruskin's father and grandfather did. Bruskin Hardware, on the corner of Fifth and Porter Streets, is still operated by cousin and locksmith Irv Bruskin.

As with the Luters, the family-owned business was handed down from father to son. Of course, one business grew into a Fortune 500 company while the other remained relatively unchanged over generations. "I remember the family joke was that the store had always been random and chaotic, but my family knew where every item was." Luter's family had a very different oft-told story, passed down to Luter by his father. A grandfather and grandson see a tombstone that reads, "Here lies a man who had no enemies." The boy is in awe, thinking the man must have been great indeed. But the grandfather scoffs: if a man has no enemies, he didn't do a damn thing with his life.[2]

Bruskin was about to become one of Luter's fiercest enemies, stealthy, cunning, challenging the kingpin by attacking his mother ship. The union activist was about to launch a national image and branding campaign that would try to bring Smithfield to its knees, or at least force it to pull up a chair at the bargaining table and spend untold millions in the process. At times over the next few years, with Bruskin at the helm of the Tar Heel campaign, it would seem as if leadership at the UFCW had taken on a bigger personality with grander dreams of success than they had ever envisioned or could completely control. Bruskin often took the bit in his teeth and ran. "By 2005, we understood that despite the importance of the NLRB decision and the strength of it, no remedy was to come out of the NLRB board," Bruskin said. "When they hired me, they knew they were done with the labor board and were going to fight a campaign forcing some kind of free and fair process for the workers. We weren't going to let Smithfield cheat again."[3]

An alternative to the NLRB was proposed in Congress in 2005, called the Employee Free Choice Act, which would require an employer to accept a union if more than 50 percent of employees submit cards stating they want a union. It is also referred to as the "card check" or "majority sign-up" section of the National Labor Relations Act. The NLRA had always allowed—but never required—a company to accept a union simply on the basis of signed cards. This new law, if passed, would create a substantial change that supporters believed would level the playing field between workers and employers.

Though they can, companies rarely choose the card-check system because the secret-ballot method allows them to fire and intimidate workers over a long period of time with impunity. Abuse of the election system had become so widespread that during a typical union campaign, one-quarter of employers would fire at least one worker who supported a union. Workers who supported the union would have a one-in-five chance of being fired.[4] In his introduction to the Employee Free Choice Act, Representative George Miller (D-Calif.), former chairman of the U.S. House Committee on Education and Labor, said that "the current process for forming unions is badly broken and so skewed in favor of those who oppose unions, that workers must literally risk their jobs in order to form a union. . . . Sadly, many employers resort to spying, threats, intimidation, harassment and other illegal activity in their campaigns to oppose unions. The penalty for illegal activity, including firing workers for engaging in protected activity, is so weak that it does little to deter law breakers."[5] Said Bruskin, "You cannot underestimate the obsession the national unions had with the Employee Free Choice Act."[6]

But Bruskin, along with many labor activists, was dubious of its success, given the intense opposition from business. If he wasn't waiting around for the NLRB, he certainly wasn't waiting around for Congress to make Smithfield accept the card-check system. Bruskin believed he could do it with his own campaign. "The question was how do you actually implement a campaign like this so it brings Smithfield to the table with the UFCW. . . . That was the challenge. . . . The strategy of just waiting for victory in the court [referring to the ongoing appeals to the NLRB ruling] wasn't going to happen."[7] The campaign, he decided, was to have three legs. "We had to campaign in the plant, we had to continue to pursue the intensive legal campaign, and we had to nationally brand Smithfield as an abusive company, which they were."[8]

---

The strategy of a corporate campaign was not a new one. It had evolved through time, building on the most successful techniques of previous generations of organizers. It relied heavily on research, community buy-in, the pressure of public opinion, economic pressure, and sometimes the courts. An early theorist was Bob Harbrant, president of the Food and Allied Service Trades Department of the AFL-CIO in the 1980s, who said that the corporate campaign created "the widest possible net."[9] Another was Jeff Fiedler, former president of the Food and Allied Service Trades Department, who became head of the Research Associates of America (RAA). In the early 1980s, Fiedler was a proponent of the strategic campaign in which the union engaged with large employers at every level. Bruskin had trained under Fiedler at the AFL-CIO, which loaned him to the UFCW specifically to take on Smithfield.

Corporate campaigns are costly and require lengthy strategizing, sometimes over a period of years. They are also a highly effective means of end-running the impotent NLRB. The union is nearly three times as likely (35 percent versus 93 percent) to win a union vote with a corporate campaign.[10] But corporate campaigns can also be very dangerous. "The full power of that employer is going to come at you," Bruskin said. "You'll be overwhelmed by their broad economic and social power. If you only focus on the local employee, it will be an unfair battle."[11]

The nature of the imbalance of power between a large employer and a group of employees trying to unionize without the backing of the NLRB is spelled out in a 1991 article titled "The Pressure Is On: Organizing without the NLRB."[12] The writer, Joe Crump, then secretary-treasurer of UFCW Local 951, pointed out that NLRB campaign wins were dropping at an alarming rate and that union membership had dropped from 35 percent of all workers in the mid-1950s to 16.1 percent in 1990. To win against an unscrupulous employer, the union needed to determine the company's primary weakness, then publicize that weakness within the company's community, including churches, suppliers and vendors, news media, investors, shareholders, politicians, employees, and the labor community. "Organizing without the NLRB means putting enough pressure on employers, costing them enough time, energy and money to either eliminate them or get them to surrender to the union," Crump explained. "If you can have an impact . . . you may find that a violently anti-union employer will decide that being fair to workers isn't such a bad way to conduct business after all."[13]

Smithfield—indeed the entire meatpacking industry—had a weakness, an Achilles heel, namely, serious safety issues. Bruskin knew that government statistics supported the claim that Smithfield had a problem, but he needed to turn it into an emotionally charged debate about worker abuse and civil rights violations. Bruskin was thinking big. He needed a masterful public relations activist to help make it happen. The campaign slogan "Justice@ Smithfield" was already in place when Bruskin took over, but he was about to make it a household term.

He hired Leila McDowell as communications director of the Justice@ Smithfield campaign. McDowell was an energetic, aggressive, and strong-minded activist, a broadcast journalist and writer, as comfortable in front of the camera as behind it. "I think I scared him the first time I came in for the interview," McDowell has said.[14] The two chatted in Bruskin's UFCW office in Washington, DC, where McDowell would end up working countless hours on the "most daunting and most fulfilling" public relations campaign of her life. "She later told me when she walked into my office and saw the mess on my desk and a picture of John Brown, the legendary abolitionist who said that slaves could not be emancipated without violence, she knew she was in," Bruskin recalled. "A strong black radical political woman like Leila might be scary to some white guys, but I had known a lot of people like Leila. We hit it off right away."

McDowell has a low, lilting radio newscaster's voice. She has worked at ABC affiliates, NPR, and CNN and as DC bureau chief for Arise TV and communications director for the NAACP. Her Facebook page is full of encouraging words of hope mixed with news of police slayings of blacks and a picture of herself with Danny Glover.[15] The former member of the Black Panther Party was no stranger to challenges, even starting her own public relations firm in a church basement in 1990 and hardscrabbling it into the national conversation. Her clients have included Amnesty International, the American Cancer Society, and Nelson Mandela. After the UFCW campaign, in March 2016, she became director of communications for Families Against Mandatory Minimums, dedicated to overturning racist prison sentencing practices. "It was the genius of her ability that made the campaign," Bruskin said. "They had numerous PR firms, God knows what they paid them, and we had Leila. It was her imagination and persistence and passion that got us the incredible amounts of coverage that had an impact."

McDowell kept the campaign to its main talking point, that the company "was brutally abusing its workers." Bruskin's team had compiled an impressive dossier of worker injuries, twice the average in meatpacking compared to other manufacturing, with an estimated 69 percent not reported. Meatpacking companies had strong incentives to underreport injuries and illnesses because businesses with fewer injuries and illnesses are less likely to be inspected by OSHA and have lower workers' compensation insurance premiums.[16]

In October of 2006, the UFCW paid the RAA—the remnants of the department at the AFL-CIO that had loaned Bruskin to the UFCW for the campaign and was paying him—to research and publish "Packaged with Abuse: Safety and Health Conditions at Smithfield Packing's Tar Heel Plant," which reported a sharp increase in the number of accidents at the plant between 2003 and 2008. The report was based on OSHA logs, medical records, and hundreds of interviews with workers as well as reviews of their compensation cases. OSHA had found multiple violations and fined Smithfield $20,175 in 2005, during its second scheduled safety inspection in thirteen years.[17]

The RAA was a nonprofit group based in Washington that had formerly been the Food and Allied Service Trades Department of the AFL-CIO. That Bruskin was on loan from the RAA and still got his paycheck from them was an "open secret," he said. The union had paid the RAA for a report that according to Smithfield contained intentional errors.[18] While the RAA held itself out to the general public as an independent research organization, Smithfield criticized it for preparing the reports—for a fee from the UFCW—that Smithfield said were false, misleading, or baseless in order to discredit the company.[19] Local newspapers were certainly aware of its ties and reported them. Both the *Virginian-Pilot* and the *Daily Press* referred to the RAA's report as union-backed when they published its findings.

The RAA reported that Smithfield had settled over $550,000 in compensation claims between 2003 and 2006 and echoed the Human Rights Watch report from 2005 calling for increased government oversight of line speed and workers' compensation claims. "Without public accountability," the report concluded, "Smithfield can continue to treat its workforce as disposable, and the company can continue to deny its employees the rights and dignities that should be afforded to all human beings both on and off the job."[20]

So McDowell had both the HRW and the RAA reports containing damning evidence. The challenge was getting the media's attention and focusing it on a rural plant far from national markets. "There wasn't a lot of interest. It wasn't a well-known company. Not like Walmart or FedEx," she recalled.[21] "How do you get coverage that shone a light on the abuses in that plant in a powerful enough way that Smithfield would care and change their misbehavior?" While she estimates she sent out hundreds of emails, the key to her success was to think like a journalist and constantly come up with new angles that were newsworthy.

McDowell's first big breakthrough was convincing Bob Herbert, then a columnist at the *New York Times*, to travel to Tar Heel and report on what he found. "Think pork," Herbert wrote in June of 2006. "Sizzling bacon and breakfast sausage." His editorial included interviews with workers, nearly all undereducated and poor, descriptions of blood and feces and workers "hollering when they're on their way to the clinic," cut by "flashing, slashing knives that slice the meat from the bones." His final salvo? "Workers at Smithfield and their families are suffering while the government dithers. . . . The defiance, greed and misplaced humanity of the merchants of misery at the apex of the Smithfield power structure are matters consumers might keep in mind as they bite into that next sizzling, succulent morsel of Smithfield pork."[22]

It was a coup. Herbert was the first of many journalists McDowell convinced. "I believe in the old tradition of advocacy journalists. The old black journalists weren't concerned about point of view. They believed that truth was on the side of justice and on the side of the people." McDowell said that she was a "crusader for truth" and that the campaign was a quintessential David and Goliath story. "On our little side was, not a little rag-tag band, but a staff of three or four in Washington. But we were relentless. I was one person up against the nine PR companies Smithfield hired. That's the power of the voices of the worker."

McDowell coordinated an advertising branding campaign that repeated the slogans of the UFCW's Justice@Smithfield campaign in every venue:

"Packaged with Abuse." "There's blood on Smithfield's products." "Calling for Justice." They hit public transportation in Washington, DC, with large, full-color banners of plant workers with quotations. One African American woman was quoted as saying, "Our managers called us the "N" word. They beat people. They broke the law and fired people illegally. . . . Smithfield cared more about its hogs than it did us. Until Smithfield does better by us, you can buy better than Smithfield." Another banner read: "Smithfield Pork: Packaged with Abuse." Still another pictured a Latino man with the quote: "Last year, I was knocked down and trampled by hogs when I worked at Smithfield Packing in Tar Heel, North Carolina. Now I can barely walk." McDowell bused in workers to events, where they set up a podium and microphone on the street under a large banner and brought their stories directly to the people.[23]

Focusing on human rights and publicizing worker mistreatment drew the ire of religious organizations, who cast the company's actions in a moral light. "America, American justice and America's principles are at a crossroads," said Dr. Bennie Mitchell, chair of labor relations for the National Baptist Convention, which had about 8.5 million members, primarily African American.[24] "Smithfield would have us go back to the dark days of company violence and threats against workers. We're here to say we won't go back. . . . God is always on the side of the oppressed, and these workers are fearful for their jobs. . . . They're treating people like automobile parts, expendable. They may be laughing at the profit table. They may be going on luxurious vacations and buying yachts off the backs of the poor. But God is not dead nor is he slumbering. He's watching, and unless these people straighten up, they will regret what they've done."[25]

---

The methods conceived by Bruskin and executed by McDowell were causing friction within the national UFCW. "Gene was the real voice in the darkness because many of the union consultants just didn't believe that consumers would care about worker abuse," McDowell said. Even in 1906, Upton Sinclair was surprised that the main result of his novel *The Jungle* was reform in federal food safety laws, not reform in improving the horrific working conditions of the workers.[26] Bruskin didn't always get approval for his strategies and declined to comment on those that were never approved. "This is a really tricky political area and there's a huge amount of controversy in terms of the campaign and how decisions were made and what the strategy was. You could say that aggressive tactics were needed and there were things not everybody was comfortable with."[27]

It's hard to imagine how daunting the challenges must have seemed to Bruskin at the time. He was managing both an internal and an external campaign, while monitoring and publicizing legal results from the slow-moving NLRB. The ground campaign—even with the steady hand of Eduardo Peña—had to accelerate to keep up with the external campaign. At a certain point, Bruskin would have to show that the majority of workers wanted a union. Whether that would end up being a card check or another secret-ballot election that included concessions from the company to protect workers' rights, Bruskin didn't yet know. "We thought card check was the best way to do it, but our backup was to force the company to agree on a fair process for the vote," he said.[28] A fair process, he believed, would include neutrality in the workplace leading up to a union vote. If the company had access to workers prior to an election, so should the union. If the company had worker addresses, so should the union. If the company could hold mandated meetings with workers, the union should be allowed to offer rebuttal evidence. And, during the vote—ideally held at a neutral setting—the company should not be allowed to intimidate workers with hired guns.

Keeping union cards updated was an enormous task for organizers in the ground campaign. The plant's workforce of five thousand people, most living within about fifty miles of the plant, was primarily rural Mexicans and African Americans in a right-to-work state, while the business owners were mainly white, with a zero tolerance for unions. Workers were divided by language and culture, and the company deliberately manipulated these differences to discourage collective action. In addition, waves of outside forces beyond their control could swell support or quickly demolish it. NAFTA had already altered the plant demographics; immigration enforcement was about to alter it again.

---

In 2003, under the Homeland Security Act, the Immigration and Naturalization Service was abolished. U.S. Immigration and Customs Enforcement

(ICE) was one of three agencies that replaced it, taking over the job of policing border control and illegal immigration to prevent terrorism.

Then, on December 16, 2005, an immigration bill—the Border Protection, Anti-Terrorism and Illegal Immigration Control Act of 2005 (H.R. 4437)—passed the House of Representatives by a vote of 239 to 182, along party lines. It sought to strengthen enforcement against undocumented workers and tighten border control. Called the Sensenbrenner Bill for its sponsor, Wisconsin Republican Jim Sensenbrenner, it unleashed a national furor, even though it failed to pass. In particular, its call to arrest undocumented workers—and penalize anyone who helped or hired them—and charge them with felonies was unprecedented. With an estimated 11 million undocumented workers working and supporting businesses across the country, fallout from the bill could have been catastrophic, according to social justice, humanitarian, and religious organizations. "It was so severe that I could be held accountable if I rented an apartment to an undocumented worker," Bruskin said. "I could barely sell a hotdog to somebody without checking their papers."

Protests began in Chicago in early March and swept across the country for the next eight weeks. There had been a small rally on April 10, 2006, in Lumberton near the plant, where workers from Perdue and Smithfield gathered in the Lowe's parking lot on Fayetteville Road. The UFCW provided signs and yellow T-shirts that said "Justice@Smithfield." "We are not criminals. We are responsible, hard-working people," said Tar Heel worker Antonio Millan, whose shirt had a hand-written slogan: "Immigrants are good workers."[29]

Even as the UFCW was trying to integrate Latino immigrant workers into Justice@Smithfield, the immigrants were pursuing their own agenda, including national efforts to improve their legal standing in the country, and that movement spread into urban and rural communities alike. As Bruskin recalled it, all hell broke loose. "A group showed up in my office and said they had decided to demonstrate on May Day and close the plant," Bruskin said. "We didn't even know some of them. These weren't the main (union organizers in the plant). . . . I said, 'What do you mean you're going to close the plant?' I was thinking to myself that we were union organizers. We've spent years trying to get people to act in solidarity and they just show up and say they're going to close the plant. So, we said, 'OK. What do you want us to do?' They said. 'Organize a march.'" Union organizers called a meeting with local leaders—a local priest, the head of a Spanish-language radio station, the owner of a local night club, soccer team sponsors, and the major Spanish-language newspaper. "We planned a march," Bruskin said.

On May Day of 2006, four thousand immigrants, according to police estimates, marched through the streets of Lumberton as part of the national "Day without Immigrants," designed to show how commerce would be affected if immigrant workers were deported.[30] The crowd wound its way through three miles of Lumberton streets to city hall and surrounding neighborhoods. "We used the fairgrounds," Bruskin recalled. "Everyone drove there with their families." Most wore white T-shirts saying, "Immigrant Rights Are Worker Rights," the local paper reported, and several carried American flags. Marchers reported that both Mountaire Farms and Smithfield Packing had to close because so many workers were absent. Lee County school officials said that half the system's two thousand Latino students missed school.

Workers condemned legislation proposing a wall on the U.S.-Mexican border and demanded a legal way to become citizens, they told local reporters. Protestors chanted a call and response:

What do we want?

Amnesty!

When do we want it?

Now!

"A people united will never be defeated," people shouted, echoing "El pueblo unido jamás será vencido," a Latin American revolutionary chant from 1973, reminiscent of "home of the brave, land of the free" in the United States. They held up signs in both English and Spanish that said: "We are not criminals"; "Yo [heart] U.S.A. Gracias"; "Love thy neighbor, America."

The mostly Latino protestors marched past Lumberton's main office buildings and through neighborhoods of mostly African Americans. Protestors pushed strollers and carried babies, Bruskin recalled, while local Lumberton families stood on the street corners as if watching a parade. UFCW supporters handbilled with leaflets bearing an image of Martin Luther King on one side and Cesar Chavez—farm-worker hero—on the other. His campaign slogan, "Sí, se puede," was adopted as Barack Obama's 2008 campaign slogan.

The workers were part of a much larger movement, with demonstrations taking place across the nation. In Los Angeles, an estimated 40,000 workers closed down about 30 percent of small downtown businesses. In Chicago, a crowd of about 400,000 gathered during lunch breaks as Irish, Polish, and Hispanic immigrants marched together.[31] But local North Carolina legislators were unmoved by the Lumberton march. Democratic U.S. Representative Mike McIntyre, who represented Lumberton, continued to support legislation to increase punishment for illegal immigrants. "We are a nation of laws, and we cannot reward illegal behavior," he told a local newspaper reporter in a telephone interview. "There is a right way and a wrong way to do things, and it's that simple." Nor were all those watching the protesters sympathetic. "I think it's a bunch of bullcrap," said Erica Floyd, 18, who was standing outside a housing project near the path of the marchers. "All our jobs done gone up and left. They're over there in Mexico."

Organizers played the national anthem at city hall, where Representative McIntyre had an office, and asked people to write Senators Richard Burr and Elizabeth Dole, asking for immigration reform that would make it easier for immigrants to get good jobs. Among the speakers was John Herrera, the first Latino elected to municipal office in North Carolina. "We want to get involved," Herrera said. "We're not only creators of culture and agricultural workers, but we are people. . . . We're producers of ideas."[32]

Lumberton, a sleepy town of about 21,000 residents, was overwhelmed. Police Chief Robert Grice said he'd never seen anything like it in his thirty-two years on the force.[33] Organizers picked up garbage and "left not a speck of dust," Bruskin recalled. Local companies—including Smithfield—contributed money toward buses to get people home and back to their cars. "It was hugely powerful and inspiring," Bruskin said. "The union had a clear role and that was a very positive signal." The rally was also a boost to Bruskin's team. "We were escalating organizing among Latino leaders," Bruskin said. Though Justice@Smithfield focused primarily on work conditions, he said that the national aspect helped broaden community support for workers.

---

Immediately after the May Day protests, the campaign got another huge boost. On May 5, 2006, in the midst of the growing vigor of the UFCW campaign, the U.S. Court of Appeals for the District of Columbia Circuit upheld the NLRB's finding that Smithfield Packing Co. repeatedly broke the law in its fight to keep out the union. The court agreed with the NLRB that the company had violated the NLRA by illegally firing ten employees who supported the union, including one who was beaten by the Tar Heel plant's private police force. The court ruled that Smithfield had threatened to freeze wages and shut down the plant, spied on union activities, and created an atmosphere that "was exceptionally hostile to union-organizing activities at the Tar Heel plant."

Key to the victory was the union beating Smithfield to the appeals court after the initial ruling. The union filed in the DC federal district court, beating Smithfield's appeal to the ultraconservative U.S. Court of Appeals for the Fourth Circuit in Richmond, Virginia. According to Judge West, in his only public comment on the case, "If the appeal had gone to the Fourth Circuit, I doubt that it would have been upheld."[34] The Fourth Circuit Court in Richmond is considered to be the "most anti-union, anti-worker appeals court in the United States, dominated by arch-conservatives," said Lance Compa, author of the HRW report.[35]

The DC appeals court—considered more liberal and sympathetic to workers—concluded that Smithfield had engaged in "intense and widespread coercion." The company was ordered to pay $1.5 million in back pay with interest to the ten workers it illegally fired. Smithfield could have appealed to the Supreme Court, but on June 15, Smithfield Packing president and COO Joseph W. Luter IV, son of CEO Joe Luter III, said that the company would not do so, despite their strong disagreement with the findings. "We have argued strenuously that the allegations the union made concerning Smithfield's conduct during both elections were false. . . . But we recognize that we have lost our case in court. When a new election is called, we will comply with the NLRB's remedies to assure a fair vote that represents the wishes of our plant's employees. . . . Smithfield plants meet the highest state and federal regulatory standards for worker safety," Luter IV said, adding that injuries were going down and that OSHA had praised the company's commitment to protecting employees.[36]

In a surprise announcement later the same day, Joe Luter III resigned as CEO, saying he would no longer personally lead the day-to-day operations of the company but would instead focus on acquisitions and long-term strategy. The announcement came after the worst quarter profits in more than seven years. While the company insisted that Luter, then 67, was not forced out, his bonus — which had been $9.8 million in 2005 — was slashed to $3.8 million.[37] Luter said one reason he stepped down was that he was tired of regulators: "It just wasn't fun anymore."[38]

That summer, Smithfield Packing was forced to offer fired workers their jobs back and pay back wages with interest to ten workers. The best part, though — for workers who had testified eight years earlier and for attorneys who had worked for years preparing the case and handling years of appeals — was that the NLRB decision would have to be read aloud in the plant in front of all the workers, in both English and Spanish.

For lead NLRB attorney Jasper Brown, the Smithfield case had been, of course, a high point in his long career. The victory was sweet indeed. "It was a great victory, a major case, one of the largest in the country and especially important in the South," Brown said. "Typically they post it, but in this case, we wanted a reading to make sure everybody got the word."[39]

The day of the reading, production was stopped as all the employees gathered in the box room, the same large, open room where the company had held its mandatory anti-union meetings. The agent stood at a same podium where Jere Null had stood ten years before. But this time, instead of threatening workers, she was reading out loud all the illegal violations of the company, including Ludlum's firing, and all the remedies — including the overturning of both prior elections and the ordering of a new election. It took the agent a long time to read the entire order, as several company managers and NLRB agents stood beside her. Then the order was posted inside the plant for sixty days.[40]

The agent read out the name of every illegally fired worker: Tara Davis, the wife of Rayshawn Ward; Rayshawn Ward; Chris Council; Margo McMillan; Ada Perry; Lawanna Johnson; George Simpson; Fred McDonald; Larry Jones; and Keith Ludlum. All received back pay. None, except Ludlum, wanted their old jobs back.[41]

For Keith Ludlum, the hog driver, who had been publicly humiliated and escorted out of the plant by armed security, it was sweet revenge. Not only did the company have to give him his old job back; they had to pay him $170,000 in back pay, covering the time when he was unemployed and the difference in earnings when he was. Ludlum said that the NLRB agent who read the order made him a star. "When they called out my name, everybody from Livestock went crazy, hollering and cheering. Everybody just kind of stopped. The lady stopped and looked at me. She was really good at being neutral. But she was really pulling for the workers. When that cheer went up, she paused for a while and let it sink in. She knew I was the only one who went back in there and she gave attention to it. It was a good victory day."[42] The back pay wasn't "life changing money," said Ludlum.[43] He added, "It's not much of a burden to the company. They'll spend that much on attorneys in a year. . . . It's just the cost of doing business."

---

There was one big problem with the appeals court's ruling. The court did not uphold Judge West's remedy that the next election be held at a neutral site. West had written, "Where, as here, an employer initiates physical violence at or near the polling place just after the election results are announced, and it engages in egregious and pervasive unfair labor practices and objectionable conduct, the reasons for favoring conducting a new election on the [employer's] premises have been substantially undermined. A new election should be conducted off premises at a neutral site."[44] The judge went so far as to note the use of Bladen County sheriff's deputies in the parking lot and suggested "a site outside of Bladen County." In the alternative, West said that employees could mail ballots.

The full NLRB board left the location of the election to the discretion of local NLRB officials. The appeals court turned down the union's request to be allowed access to the plant. And it did not mention enforcing an off-site election. This meant the union would have to delay a secret-ballot election as long as possible to build up a strong majority in the workforce, strong enough to withstand another round of company harassment and intimidation of workers, actions Bruskin believed would once again break labor laws and start the protracted NLRB process all over again. He had no tolerance for that losing scenario; he wanted to win, as swiftly as possible, either through a card-check system or by forcing Smithfield to agree to neutrality provisions during an election, including instant remedies for alleged violations of fair labor practices—both leading up to and during the vote—that would completely remove Smithfield's tool of delayed justice.

Initially, the company—though required to reinstate Ludlum to his old job driving hogs—assigned him to hog tattooing in the same division. This isolated him from his co-workers and kept him from organizing. "The job was extremely boring, filthy and tedious," Ludlum said.[45] "They wanted to keep me quiet and make my life miserable." The NLRB and the union stepped in, and the company moved Ludlum back to driving hogs.[46]

At first, Ludlum was low-key, wearing his union shirt and signing people up during breaks, as usual. He wanted to size up the situation, analyze relationships, figure out who was talking to whom, and who was reporting to management. But, as time went on, he grew bolder. After being publicly exonerated and apologized to in two languages, Ludlum was fearless. When he was encouraging workers to sign their union cards and management told him to leave the property, he would openly confront them. "Are you firing me?" Ludlum would ask, smiling. "Because they just told me in court you can't tell me to leave the property unless you're firing me." He said that the looks on the supervisors' faces were priceless. "They were used to saying whatever they wanted."

After work, Ludlum would stand in the main hallway of the plant, handing out union literature. "It was quite a feat," he recalled. "You have to understand it's hard for people to stand up against their employer. They hand you a paycheck every week and you're going to stand up to them?" One day a supervisor came into the hallway and asked for Ludlum to leave. Ludlum asked him for his name and position. "I told him, whoever sent you, tell them we're not going anywhere. Under federal law, we are allowed to be here."

In July of 2006, close to a thousand black and Latino workers came to work at the Tar Heel plant wearing Justice@Smithfield T-shirts, declaring it "yellow fever" day. "We linked the denial of basic human rights of these Latino workers with the African American community," Bruskin said. "It showed the African American community why we needed to work together."[47] With Ludlum a key insider for the union, working the local ground game at the plant, Gene Bruskin was once again turning his attention toward the economic pressure campaign against Smithfield.

## CHAPTER 13

# Pressure Mounts on Harris Teeter
# and Paula Deen

arris Teeter was Smithfield's single largest customer and the largest supermarket chain in North Carolina. In the 2000s, the upscale retailer wanted to break ahead of competitors with newly "branded" departments, including "Harris Teeter Rancher" beef in June 2002 and then bacon from Smithfield's Tar Heel plant. In the summer of 2006, UFCW's Justice@Smithfield began a synchronized protest campaign with a full-time community organizer in North Carolina coordinating church and civil rights leaders who literally cried out against the worker abuse at the Tar Heel plant. By the time they were done, the phrase "packaged with abuse" became synonymous with Smithfield Foods. "We never told people to boycott — never used that word — my idea in terms of strategy," said Gene Bruskin. "But when you stand in front of supermarkets and hand out flyers that tell people to complain to the manager, people just don't buy the product. We were putting pressure on Harris Teeter stores to get rid of the product."[1]

The campaign widely publicized its "Day of Action" events at stores in Asheville, Charlotte, Durham, Fayetteville, Greensboro, Hickory, High Point, Raleigh, Rocky Mount, Wilmington, and Winston-Salem. Flyers documented the NLRB rulings about violence, threats, and intimidation of workers. It also included information from HRW (Human Rights Watch) and the RAA publications about dangers in the workplace, injury rates, and denial of compensation claims. Organizers cited congressional testimony about the company's deliberate inciting of racial tension and segregation in the workplace.

"The most significant thing we did was joining the black community in North Carolina into a campaign not to buy Smithfield items from Harris Teeter," said the Reverend Nelson Johnson, co-founder of the Southern Faith,

Reverend Nelson Johnson, co-founder of the Southern Faith, Labor and Community Alliance. Photo by Raul R. Rubiera, *Fayetteville Observer*, August 18, 2005.

Labor and Community Alliance in Greensboro. "We had weekly phone calls from clergy and community activists from Hickory across the state to Greensboro to Raleigh to Durham and other places. The idea grew from my discussion with the pastors, with the NAACP, and community leaders."[2]

Working together, religious leaders, civil rights groups, and consumer advocates used methods gleaned from a century of labor-organizing tactics: they chanted; they handbilled; they called on workers to give speeches about their mistreatment. "Central to the entire campaign was giving workers a chance to tell their own stories and giving people the chance to meet them," said Bruskin. "That's what was so convincing. It wasn't some leaflet or my persuasions. We took workers and when people heard them, they literally were crying."

On July 9, 2006, in front of a Harris Teeter store on West Market Street in Greensboro, more than forty protesters, wearing yellow T-shirts with the slogan Justice@Smithfield, chanted and waved signs about working conditions at Tar Heel.[3] On July 21, 2006, according to the local paper, about thirty people gathered in front of a Harris Teeter store near Raeford Road in Fayetteville, wearing yellow T-shirts and waving signs that said "Honk for Justice"; "Where does your BBQ come from?"; and "Smithfield is guilty." A union intern called through a bullhorn, "When I say Smithfield, you say sweatshop." When a delegation went inside to talk with the manager, one person returned a Smithfield pork tenderloin. The manager was polite and reimbursed him.[4] On December 2, as part of a statewide Day of Action, Reverend Joe Brown, an African Methodist Episcopal (AME) minister, called out to a crowd of about thirty protestors in Wilmington, "May our Christianity not just cause us to sit in pews!" Protestors held signs reading "Smithfield Bacon: Packaged with Worker Abuse."[5] The protest was one of eleven simultaneous protests across the state.

Johnson's Southern Faith, Labor and Community Alliance, a coalition of state faith-based and community organizations, organized the protest together with the UFCW and sent a letter of protest to Harris Teeter. An NAACP vice president from Wilmington said he asked a manager to consult his conscience, but the manager refused to speak with protestors or comment. Leadership urged churches to get involved.

The protests were condemned by Smithfield's director of corporate communications, Dennis Pittman, who wrote a letter to the editor of the *News & Observer* that ran on December 6, 2006. Pittman complained that a news story about a Harris Teeter protest was not fair in its presentation and that information circulated at the rallies was false. But the company was singing the wrong song to the wrong tune. Pittman apparently believed that consumers were making the link between union votes at the plant and the protests at Harris Teeter. They were not. So when he directed his comments to the union dispute in Tar Heel and failed to address the mistreatment of workers, his message didn't land. "Smithfield has offered to schedule another new secret-ballot election at any time," Pittman said. "We have even offered to pay half the cost of an independent outside observer to ensure the vote's fairness. But the union has rejected our offer. Why? If the union truly believes that employees want a union, it should schedule a secret-ballot election."

Bruskin was not about to agree to another election without neutrality agreements. And the public didn't seem to care. They did not associate the protests with the UFCW's labor organization campaign. They associated the protests with Smithfield's abusive treatment of its workers. The campaign had taken on a life of its own, and Smithfield was losing its ability to control the message or to contain it—as it had in the past—by appealing to southerners' distaste for unions.

By the end of March 2007, the campaign had spread into a Southeast Day of Action. On March 31, weekend protestors showed up at twenty-two stores in eight states, urging Harris Teeter to remove all Smithfield Packing products. "Smithfield is not listening, so we're going to have to go after their pocketbook," Leila McDowell told the Associated Press, whose story went out on the wire to about 1,400 newspapers and thousands of radio stations. Protestors rallied outside Harris Teeter stores in Asheville, Chapel Hill, Charlotte, Durham, Fayetteville, Greensboro, Hickory, High Point, Raleigh, Rocky Mount, Wilmington, and Winston-Salem in North Carolina; Arlington in Virginia; Nashville in Tennessee; and Charleston and Florence in South Carolina. "We are telling Harris Teeter we don't want pork that's packaged with worker abuse," said the Reverend Nelson Johnson. Other organizations that joined the protests included the Northern Virginia Labor Council, the Chapter of the Association of Community Organizations for Reform Now (ACORN), and Interfaith Worker Justice.[6]

Smithfield, still singing the wrong tune, tried to fight back. "If employees want a union, we welcome an election at any time," Smithfield's spokesman Dennis Pittman said in response to protesters, adding that he had no concerns about protests planned in North Carolina, South Carolina, Virginia, Tennessee, New York, Massachusetts, Illinois, and Indiana for that Saturday. Perhaps he should have, because Smithfield was getting a lot of bad press, and that was leaving a bad taste in the mouths of consumers. The union's smithfieldjustice.com website and flyers included product coding to help consumers identify products manufactured in the Tar Heel plant.[7] On March 28, the UFCW sent out a press release that they were "encouraged by the progress we have seen Harris Teeter make in removing Smithfield Tar Heel pork." Smithfield said that wasn't true.[8]

---

The Harris Teeter campaign peaked on June 17, 2007, with a Father's Day protest at the home of the grocery chain's president, Frederick J. Morganthall II, in Charlotte. Morganthall, who favored striped ties and starched white shirts, had been with the chain for nearly thirty years, working his way up to president by 1997. A 1973 graduate of Bowling Green State University with a BS in journalism, he had made it publicly known he was against government interference in the grocery industry. In 2006, he earned just under $1 million per year. At the time of the protest, he was earning $1.25 million in total compensation.[9]

"Fred," then 56, lived with his family on Stonecroft Park Drive in Charlotte when the union dropped by for a visit, along with a few friends. The light terra cotta stucco–style two-story house was immaculately groomed, with a sweeping lawn and broad steps. The 4,840-square-foot, five-bedroom, six-bath home was worth nearly $1 million.[10] The Olde Providence South neighborhood included a racquet club, premier tennis club, country club, and Olympic-sized swimming pool.

"What we did, we took a couple busloads of workers and community supporters and their children and workers' kids in buses from Tar Heel to Charlotte," Bruskin said.[11] The group of about two hundred protesters held a rally at Harris Teeter, then marched to Morganthall's house. "Reverend Nelson Johnson helped coordinate the march. . . . There were a lot of children. . . . It was Father's Day. They carried signs saying, 'Be Nice to My Daddy' and 'Don't Treat My Daddy Bad,' that kind of thing."

As Bruskin recalled, the police escorted the group into a gated community of "really huge beautiful homes." The children climbed the stairs to Morganthall's home carrying a large Father's Day card. "They knocked on his door, but nobody answered. Whether he knew we were coming, I don't know." Bruskin mused that the large number of poor African Americans and Latinos in their neighborhood had a "lot of people looking out their windows." Bruskin couldn't help but note the difference between the families, some of whom lived in run-down trailer parks, and those in the gated community. "The children walked up this long flight of stairs, in this impeccably kept community of stone houses, up to the door. It was immaculate. They rang the bell and had a big card, signed as a poster, saying 'Don't Forget Our Fathers.'"

After about twenty kids had actually walked to the door, Bruskin recalled, someone had the idea to try to call Fred. "We had everyone take their cell phones out and call simultaneously and two people got through. . . . He picked up. We said, 'Hi Fred. We're in front of your house with two hundred people and kids to say Happy Father's Day.' He was stammering, 'Why are you bothering me? I have nothing to do with this. I just sell the products. It's not my fault.'"

Soon after that, Smithfield's Tar Heel products began disappearing from Harris Teeter stores. "It was a direct result of community delegations going into stores, asking to meet with the manager, handing over a packet of materials about worker abuse, and asking them to stop," Bruskin said. "This was the culmination. This was an extremely conservative, white, evangelical-run company. They would sooner die than be connected to abuse." Bruskin is still amazed that the protestors weren't stopped. "We fully expected to be blocked. Of course, we were peaceful. A minister was there. We reassured the police there would be no trouble." The UFCW Smithfieldjustice.com website posted news about the event called "Prayers for Our Papas," which said, "Workers at Smithfield Packing in Tar Heel are abused."

While Harris Teeter never publicly stated it had pulled Smithfield products, the union systematically monitored the presence of products in Harris Teeter stores, reporting a near total removal of the product. The entire process was extremely upsetting for Morganthall. He repeatedly contacted Smithfield and asked them to keep the union away from his stores.

---

The boycott crept up the East Coast. In December 2006, organizers protested outside Johnnie's Foodmaster in Somerville, Massachusetts. UFCW publicity around the event said Smithfield was "implicated in worker cruelty" and that workers "face degrading and dangerous working conditions."[12] The following August, protests were held outside grocery stores in Boston. In February of 2007, protests sprang up in Ann Arbor, Michigan, then in Atlanta in June and Nashville in July.

In a New York City council meeting on October 25, 2006, Resolution 582-06 called "on the City of New York to cease purchasing products from Tar Heel, North Carolina's Smithfield Packing Company, and urged supermarkets operating in our city to cease purchasing Smithfield products from Tar Heel, North Carolina." The resolution was referred to the council's Committee on Civil Rights for further consideration.

On May 24, 2007, the United Church of Christ published a "Resolution on Worker Justice at Smithfield." On June 4, 2007, the Cambridge City Council published Policy Order Resolution O-17, condemning Smithfield's alleged workplace practices at the Tar Heel facility, and Mayor Joseph Curtatone, of Somerville, Massachusetts, proposed a resolution "to ban purchases from the Smithfield Packing Co.'s Tar Heel Division." On June 25, 2007, the Chelsea City Council published a resolution condemning Smithfield for its workplace practices, encouraging "all supermarkets and vendors in Chelsea to boycott the sales of Smithfield meat products in their stores."[13]

On August 1, 2007, the Boston City Council resolved that the City of Boston should "review its purchasing of any products from the Smithfield Packing Company in Tar Heel, North Carolina . . . and suspend these purchases." On September 26, Thomas Menino, the mayor of Boston, wrote a letter to C. Larry Pope, CEO of Smithfield Foods, and Joseph Luter III, chairman of the board of Smithfield Foods, in which he told them of the resolution and expressed "serious concerns regarding the recent violation of federal labor laws by Smithfield Foods, Inc." Menino informed the Smithfield executives that the city would continue encouraging Boston-area supermarkets to suspend offering Smithfield products until the "workers for Smithfield Foods, Inc. can work in a safe environment with dignity and respect."[14]

Every step of the way, UFCW publicist McDowell was bending the ear of journalists, getting publicity for the protests and resolutions. "We were causing major internal problems in the company," Bruskin said. With the combination of the damning report from the NLRB judge in 2000, the publication of the scathing HRW report in 2005, the 2006 RAA report containing gruesome details of individual worker's stories, the successful removal of products from Harris Teeter, and resolutions citing worker mistreatment in major urban markets, Smithfield found itself barraged on all sides.

The campaign had ceased to be about union organizing and was growing into a blistering human rights issue, with the union painted as the protector of those human rights. The company's continued attempts to focus the topic on unionized labor and calling for a new union vote were not working; they watched helplessly as the boycott expanded from a regional to a national platform. The union was successfully dodging the typical southern aversion to unions by changing the focus.

On April 3, 2007, the *Fayetteville Observer* published an editorial written by Gail D. McAfee, a minister who served on the board of directors for the Southern Conference, as chairwoman of the Eastern North Carolina Ministries Council of the United Church of Christ, and as a spokesperson for the

Ministers for Racial, Social, and Economic Justice. She made the case that a union would protect the workers and reduce the rates of injury. A union, she argued, would end worker abuse. "A human tragedy is currently unfolding in Tar Heel. Imagine going to work in the morning and not knowing if you'll still have your 10 fingers when you get off work. . . . That's the testimony of an injured worker from Smithfield Packing's Tar Heel facility. . . . Smithfield Packing has been found to have unlawfully assaulted, fired, intimidated, used racial epithets against and threatened its workers with violence and arrest . . . to silence the over 5,500 workers at its flagship facility and prevent them from obtaining the protection of a union contract. . . . As a concerned people, as Christian people, it's time to act . . . now! . . . We call for justice at Smithfield now!" She cited statistics showing that between 2003 and 2006, according to the Occupational Safety and Health Administration, injury rates at one of Smithfield's union plants had dropped by half, while at Smithfield's non-union plant in Tar Heel, injuries had doubled.[15]

---

By July, the NAACP was considering launching a national boycott. On July 19, Reverend Dr. William Barber II joined approximately one hundred protestors in front of the plant chanting and holding signs. They said they were demanding justice, and Barber announced that the NAACP was going to respond on a national level. Two workers said they had been fired when hogs fell on them while working. Another said the company did not provide adequate drinking water to workers in Livestock. "This is serious business," Barber said. "We intend to push forward."[16] He also said, "The NAACP is a connected organization. Whatever the position of the state and national leaders, that is the position of the local branches. . . . When the NAACP makes a decision, everyone knows."[17]

The NAACP announcement came as the union began a limited television campaign featuring workers who had been illegally fired during the 1994 and 1997 union campaigns. Keith Ludlum, the hog driver fired during the first campaign, was featured. "I was fired illegally for my union support. My wife was pregnant at the time. It took me twelve years fighting Smithfield in court to get my job back."[18]

On August 2, Barber announced a boycott might begin unless Smithfield agreed to a "fair and reasonable meeting with the union, workers and a third

Reverend Dr. William Barber II, president of the North Carolina chapter of the NAACP, speaking to a crowd of protesters in front of the Smithfield Packing plant on July 19, 2007. *Fayetteville Observer*, July 19, 2007.

party. If they refuse, then we could call for a national boycott." When the company refused, the UFCW issued a nationwide directive to local unions to begin a national campaign to pressure stores into dropping Smithfield products.[19] Local union shops were told to contact stores to let them know if they sold them, then there would be protests. On August 29, a union official in Mineola, New York, sent a letter to King Kullen, a retail food store, stating that the union would leaflet customers if Smithfield products continued to be sold.[20] The UFCW's Justice@Smithfield website provided a letter consumers could send to store managers asking them to remove Smithfield products from the Tar Heel plant.[21] It said: "Smithfield Packing Co. uses its employees as if they were disposable. . . . [It] also has intimidated its workforce though racial epithets and threats of termination for demanding basic rights on the job such as health and safety, and dignity and respect on the job." The letter went on to ask the grocer to refuse to carry Smithfield products and provided a code indicating whether they were processed in Tar Heel.

Smithfield tried to combat the campaign with its own set of television ads featuring workers pleased with safety and wages. The company said that the ads—which ran in English and Spanish—were a good recruiting tool. The battle, reported the *Fayetteville Observer*, "has escalated into a nationwide image war, with both sides relying on publicity campaigns, staged rallies and increasingly biting accusations."[22] Remarkably, Smithfield was still not willing to take a seat at the negotiating table with the union. Now the second phase of the economic pressure war, the classic corporate campaign, was about to begin. It was going to be tricky, because it targeted the widely popular—and noncontroversial—Paula Deen.

---

Deen had signed a lucrative endorsement deal to promote Smithfield hams in September of 2006. Smithfield listed the partnership first in its high achievements for that year and by 2008 was selling 2 million pounds of Paula Deen–brand spiral hams. Deen agreed to use her "folksy charm" to promote Smithfield in recipes and wore a black apron with a white Smithfield logo in photographs and appearances. Her smiling face had become synonymous with Smithfield Foods and appeared on the label of her own line of hams, the Paula Deen Collection, including the Smithfield Crunchy Glaze Spiral Sliced Ham. The deal was the perfect marriage—between America's "Butter Queen" and the corporate "king of pork."[23] (Deen was known for using large amounts of butter in her recipes.) "She was a national celebrity," Bruskin said.[24] "She was emerging with her show, books, and magazines, her restaurants in Savannah. Tourists by the busloads were showing up."

Deen's personal story was a sad yet miraculous one, which she told with gusto on television and radio stations, laughing loudly. Born in Albany, Georgia, in 1947, Deen developed debilitating phobias and depression. Despite this, she started a catering business, The Bag Lady, out of her home, then went on to open a restaurant, The Lady and Sons, in downtown Savannah. A successful cookbook propelled her onto the QVC shopping channel, and in 2002 Deen landed her own show, *Paula's Home Cooking*, on the Food Network. In 2005, Deen launched her own magazine, then published three cookbooks and a memoir. In June 2007, she won a Daytime Emmy for Outstanding Lifestyle Host.

"Her narrative was she had a tough life and had phobias about going outside, but she had overcome them all," Bruskin said. "So, she had this success story. And then she had her cooking. It's been said if she put butter on an aspirin, it would taste good. Her food was so unhealthy, she probably single-handedly increased heart attack rates during the high point in her career."[25] In the spring of 2007, Deen appeared to have been blindsided when union supporters began to protest during her public appearances and book tours. An ever-growing solidarity group of religious and civil rights groups appeared wherever Deen did, much to her growing consternation. "At a book-store, people would get in line, then unroll a scroll with all the workers' names on it until the managers would pull them out. Five minutes later, more people got in line. They couldn't tell which ones were us," Bruskin recalled.[26]

Public appearances were just as bad. "We would handbill outside," Bruskin said, "then inside, people in the balcony would drop a banner or walk up to the stage and ask her to please help." During a sold-out book tour in New York City in April, about two dozen people held a prayer vigil outside the Museum of Natural History. As she responded to audience questions, a union member tried to speak and deliver a letter.[27] "Now, I'm not an expert on the union situation but here's what I do know: I know the folks at Smith-field care about their employees and work hard to support the communities where they live, work and raise their families," Deen said afterward.[28]

The North Carolina Council of Churches wrote Deen a letter describing conditions at the plant, including information based on OSHA injury reports. Protestors showed up at the CenturyTel Center in Shreveport, Louisiana, on June 2, 2007, where shows featured cooking demonstrations, tips, and a question-and-answer session with Deen.

In Charlotte, North Carolina, on June 30, 2007, as hundreds waited in line, about sixty protestors kept them company "armed with loudspeakers, signs and a papier-mâché pig with a menacing sneer."[29] The protesters, who gathered on a sidewalk across the street from Ovens Auditorium, chanted "Hey hey! Ho ho! Smithfield pork has got to go!" and carried signs saying, "Smithfield bacon: Packaged by worker abuse." Protesting workers talked with reporters. "They treat the hogs better than human life," said Vincent Nash, 40, of Fayetteville, who had worked at the plant for two years. "It's ridiculous. I want the world to know this." The city of Charlotte was part of a national tour that included Nashville, Dallas, and Atlanta. Protestors were bused in at every opportunity. "This was an opportunity for us to bring the campaign into places we never thought possible," Bruskin said.

But the campaign had to be careful. Deen was loved by her fans, so Bruskin and his team came up with the idea of acting as if she must not know what was happening to workers and that maybe, if she knew, she would help. At one Kentucky event, Bruskin recalled there were about five hundred people paying $40 to $50 to hear her story. "She was so popular, we didn't want to slam her. People adored her. So we wrote her a letter with a group of workers to sign it, then blew it up on a four by eight sign, basically saying, 'Paula Deen, Won't You Help Us?' Our message was, 'we know you're a good person. If you knew how badly we were being treated, you'd please tell them to stop.' This was a huge issue for Paula Deen. She was a public figure and she couldn't hide."

Several high-profile journalists asked hard questions of Deen during routine cooking interviews, expanding publicity for the campaign. In August of 2007, Larry King confronted Deen on his show, asking about safety issues at the plant and why she hadn't met with workers at the plant as the union had requested. Deen protested, saying that she had met with workers and that the workers had voted against the union. "I've been to the plant. I've met with workers with no one around but me and the workers," Deen said. "Smithfield is a good company. It's a good company. And these people are so happy to have their jobs, good benefits, and you know. They're happy. . . . They're happy." King offered to host a meeting with workers on his show.[30]

Similarly, on November 28, on the Diane Rehm show, distributed by NPR and broadcast on WAMU 88.5, Deen was promoting her new book, *Christmas with Paula Deen: Recipes and Stories from My Favorite Holiday.* For about forty minutes, Deen told her story while Rehm praised her, telling listeners, "You can hear her sweet laughter." Deen said that she became agoraphobic after her mother and father died one after the other when Deen was in her late teens and early twenties. She was miraculously cured in Savannah after spending two months in bed crying. "The serenity prayer came to me," she said, mixing up and stumbling over the lines of the prayer as she recited it. "I got out of bed, honey."[31]

Unfortunately for Deen, McDowell had found out about her appearance and had red-alerted, via email, her ever-growing constituency. "She was a legitimate target," McDowell said. "She was the voice for a company harming workers. We called in and asked questions."[32] McDowell called the producer and said that "it would be unconscionable to have this woman on and not ask her these questions."

During the show, Deborah from Greensboro called in to say she was concerned about Deen's attitude toward workers in Tar Heel. "They didn't vote the union in because they were terrified. People have died. People have been hurt. Women have had miscarriages," Deborah said. "I wish you'd be more compassionate and sit down with the workers." Deen said that she had met with workers. "They absolutely are not interested in the union. . . . I can only judge by people I've met with and accept their words as the truth, that they love their jobs. They have good compensation plans."

With six minutes remaining in the show, Diane Rehm told Deen, "I have to tell you. We've had numerous emails about the Smithfield situation and other callers who wanted to talk about this, but clearly you find yourself in an awkward position." Deen agreed. "Absolutely it's awkward. I have no knowledge. This is not my expertise. I wouldn't even want to try to [stammers] convey the thought I'm an expert. I'm expert in the taste of good ham."

Rehm went to an email from Alexandra with the National Farm Workers Ministry. "You're obviously compassionate and loving," Alexandra wrote. "I'm disappointed you associate with Smithfield. People are suffering. I hate to think you're turning a blind eye. Please use your position in the public eye to stand up for the men and women." Rehm asked Deen, "Do you think there's something you can do as a public figure to speak out on their behalf?" Deen replied that it was hard. "The people I've met and talked with. They aired no grievance. This is work, not play. They like their jobs."[33]

---

The campaign ratcheted up the pressure. That December, Deen had an appearance in Chicago. Outside, about two to three hundred protestors chanted for her to do the right thing. They handed out brochures to those waiting in line with a message that appeared to be from Deen. The outside of the twice-folded piece of paper showed Deen's smiling face and said "Paula Deen Welcomes You." But inside was a dancing cartoon pig wearing a chef's hat, saying "Justice@Smithfield: Paula, Have a Heart." It went on: "Today, Paula Deen is here to share her down-home charm and tips for country cooking . . . [but] we fear she is unaware of the brutal working conditions at Smithfield Foods' largest pork processing plant in Tar Heel, N.C. . . . Paula is a paid spokesperson for Smithfield Foods. Smithfield workers have asked to speak to Paula about the harsh working conditions at the Tar Heel plant. While on the Larry King show Aug 1st, she said she would, but Paula has *not* kept her promise. 'Paula, Keep Your Promise!' Meet the workers and learn why the Tar Heel products you promote are PACKAGED WITH ABUSE!"

People waiting in line don't have a lot to do, and the flyer's designers had taken that into account. When readers opened the folded paper all the way, there was a full-page letter, dated April 15, 2007, to Deen from Tar Heel workers. "Workers have suffered terrible injuries at Smithfield. Some of us have had our fingers cut off; some of us have serious repetitive stress injuries, knife wounds, broken bones, and severe infections. . . . Sometimes we are fired if we get badly injured and then we are faced with high, ongoing medical bills and the inability to support our families. . . . We believe that you are a woman of conscience. . . . We were surprised . . . that you were helping sell Smithfield products but we believe it's only because you don't know the truth. . . . We have faith that once you know the truth you will not promote a product that is made through the pain and suffering of our families." Needless to say, word of mouth in a line of people goes a long way.

The campaign's next move was swift and very costly to Smithfield. Deen was to appear on *The Oprah Winfrey Show* the week before Christmas with best recipes for Christmas Day, featuring a Smithfield Ham. Sales were expected to spike during and after her appearance. "The Smithfield plant in Florida was producing hams and stacking them in a warehouse getting ready for her to go on TV," Bruskin said. "They were expecting to ship out huge numbers of hams all over the country."

"We jumped into action," McDowell said, "Rapid response." Eventually she tracked down someone who had Oprah's ear. "Producers don't want to be called, so it took a lot of digging," she said. "It was pretty frantic. Emergency. Emergency. I went through five or six people saying, 'I don't think Oprah would want to be involved with this.' Then I emailed information. Then we waited. We were on pins and needles about whether it worked."

The context presented to Oprah, said Bruskin, was that the UFCW was trying to save her from a huge political fiasco. "Look, we're big fans of Oprah and you need to let her know that having Smithfield Foods featured on her show is going to put her in the middle of the most contentious labor struggle in the country." Oprah responded swiftly, letting Deen know she could come on the show, as long as she didn't mention Smithfield Foods or bring a Smithfield ham. "Those hams sat in the warehouse of the company,"

Bruskin said. "Smithfield ran into court. They'd already sued us, but tried to get a judge to enjoin us. They claim they lost 900 zillion dollars, but we didn't do anything wrong. . . . We just saved Oprah."

According to the company, the union "deprived Smithfield of an incomparable marketing opportunity" by persuading Oprah Winfrey . . . to withdraw an offer to allow Smithfield's principal spokesperson, celebrity chef Paula Deen, to promote Smithfield's product before millions of viewers."[34] Said one of Smithfield's lawyers, "It's economic warfare." A union lawyer responded, "It's traditional free speech." Or maybe it was just good old-fashioned politics.

Oprah had just come off the road after stumping for Barack Obama, pulling in 30,000 supporters on December 8 at Williams-Brice Stadium in Columbia, South Carolina, the largest crowd at that point in the 2008 presidential race.[35] The UFCW's PAC had spent $673,309 in independent expenditures promoting the election of Barack Obama and would contribute $1.8 million on behalf of Democratic candidates at the national level in 2008.[36] On February 14, 2008, *Mother Jones* magazine reported that the SEIU was endorsing Obama, pledging support from nearly 2 million members, more than any other union in the United States. The UFCW endorsed Obama the same day.[37] Perhaps Oprah knew the power of the UFCW and where Obama's sympathies would lie.

The union campaign was brutal and costly, both to Smithfield's reputation and to its bottom line. It was good old-fashioned hard-ball politics, the kind increasingly gentrified unions were reluctant to use. Behind closed doors, Bruskin had won the approval of high-ranking union officials for his high-stakes campaign, but the retaliation that was to follow from Smithfield was both swift and unexpected.

## CHAPTER 14

# Latino Workers Walk Off the Job;
## ICE Raids Cause Flight

**W**hile Bruskin directed the public external pressure campaign and Eduardo Peña managed the outreach to workers in the field, Keith Ludlum was becoming the center of union strength inside the plant. Two other Livestock workers, Terry Slaughter and Oliver "Ollie" Hunt—one black and one Native American—were prime recruiters as well. "We were like co-conspirators. We were tight. I would break out my ideas to them and sell them to the rest of Livestock," Ludlum said.[1]

It took a while to build up a crew, Ludlum recalled, and he sought legal help and guidance almost daily from Bruskin and his staff in Washington, discussing liabilities and how far to push his role. But ultimately he did whatever he wanted. He was a big man with a big personality, a loud voice, and a hot temper. "I made both sides nervous," Ludlum recalled. "I was a wild card. I could do what I wanted. I hadn't cut any deals with the union."

According to Ludlum, union support was still weak inside the plant. He began to realize that some union supporters who might talk a good game in their living rooms weren't up for a public fight inside the plant. Ludlum decided to show them how it was done, to build strength by showcasing victory on the floor of the plant. He had done his homework. Anything general employees were allowed to do, union supporters could do. If someone handed out flyers for the Boy Scouts or a church picnic, then a union supporter could do the same; if another employee used company paper to print out a letter, a union supporter could do it too. Plant workers routinely put stickers on their hard hats or wrote the name of their favorite sports team with markers. Ludlum had documented the dates and times in his journal. "One of the most powerful weapons is pen and paper. You document every day, no matter how slight. It will accumulate into an effective tool," he explained.

Once he had enough evidence to be protected under the law, Ludlum used a marker to write "Union Time" in big letters on his hard hat. When morning management pulled him out from running the hogs and ordered him to take off his helmet, he was ready. "You can't keep me from writing on my hat when you've let other workers write on theirs," he told them in front of his co-workers. To avoid a fight, Ludlum handed over his hard hat, but that night he wrote "Union Time" on the back of his big yellow raincoat. "I gave you my hard hat, but this jacket ain't coming off," he told supervisors the next day. "You'll have to fire me to get this jacket off my back in this plant." His co-workers gathered around as he explained that it was freedom of speech. Four workers gave him their jackets after work to letter that night. "I pimped them up. I took metallic markers and wrote all kinds of slogans."

Livestock was now the new hotbed of union activity. "When you see someone dance against their employer and win every day, it changes people," Ludlum said. After the union put legal pressure on Smithfield, plant manager Larry Johnson wrote a letter to all the workers stating that the company did not have a problem with slogans written on hard hats and that supervisors had not been acting at the direction of management when they ordered workers to remove them. Dated June 27, 2008, the letter was in English and Spanish: "We do NOT have a policy that forbids union related stickers or writing on hard hats. . . . We apologize for this miscommunication."[2]

Ludlum won the first round. Hundreds of workers began wearing the slogan "Union Time" on their helmets and clothing.[3] "It was a huge victory for me. Here's plant management in a letter saying I was right. There were five thousand copies of that letter sent out. I was busting them wide open." The letter changed the whole dynamic in the plant. Ludlum would ask for a worker to sign a union card and hand him or her a copy of the letter. "This is why they called me Smithfield's worst nightmare," he said.[4] The union was concerned that Ludlum was pushing too hard, too fast. They wanted him to cool it. But Ludlum had educated himself on labor law during his absence. It had taken a long, long time, but he was empowered. "You lose the workers by losing to management. I was walking that tightrope. I had a crew and nobody could touch them. And nobody was going to talk to us alone because somebody was always going to write down everything you said. Even Livestock management was hands off us." Ludlum described an atmosphere on the plant floor that was worlds away from the one he had left in 1994. While

Peña concentrated on organizing Latino workers, Ludlum concentrated on African Americans, first winning over Livestock, then taking his message to the kill floor and beyond.

Whereas the company had been careless during the first two union votes, now it was being very, very careful. Between the damning NLRB decisions being upheld, the damaging national PR campaign and costly Harris Teeter boycotts, and Paula Deen's growing panic, Smithfield Foods was on the defensive. "The stars just aligned," Ludlum recalled. Ludlum felt the groundswell of union support, but he was becoming agitated and restless for something bigger, something more dramatic. "It was incredibly frustrating. I wanted something to happen. I wasn't working in shit for nothing."[5]

It may have appeared that the stars were aligning for the union, but behind the scenes Smithfield was under attack from multiple sources, including tremendous pressure from immigration officials. In June of 2006, unbeknownst to union officials or workers, Smithfield had secretly agreed to participate in a new ICE program called the Mutual Agreement between Government and Employers program (IMAGE) to quietly help identify undocumented immigrants within its Latino workforce, which had grown to more than half of its five thousand workers.

By October, Smithfield had identified about 20 percent of its workforce as having problematic paperwork—meaning primarily that the Social Security numbers they had provided when hired didn't match the government's database—and gradually sent out a thousand "no match" letters to employees giving them two weeks to prove their paperwork was valid. "On November 13, over thirty were escorted out of the plant," recalled Peña.[6] By mid-November, about seventy-five workers had been fired because their names did not match their Social Security number, and another six hundred had unverifiable identification. At the beginning of each week, the company was quietly hiring about one hundred African Americans to take the place of disappearing Latino workers.

On November 8, Pedro Mendez—who had worked at the plant for nine years—was called into HR. According to Mendez, when he asked for a copy of the no-match documentation, Smithfield would not supply it.[7] "It was agonizing," one organizer recalled. "About thirty people would get letters,

then the next day another thirty. It was a snowball effect psychologically."[8]
On Monday, November 14, entire groups of workers didn't show up for work
at all because they couldn't provide documentation within the two-week
deadline.

Management tried to be discreet, so as not to spook the workers. At different
points during the day, Latinos were told to report to HR and then didn't
return to their work stations. By day's end, their families would realize they
were gone. By that Wednesday, the workforce was in a panic. People came
to work, then disappeared. Bruskin told Peña to schedule a meeting with
workers for Thursday night.

But the workers weren't waiting for the union. On the morning of Thursday, November 16, 2006, dozens of workers simply set down their knives and
walked out, shutting down processing lines. Initially, they headed for the
smoking cafeteria, but it was soon too crowded. "We decided to move outside
because there wasn't enough room for all of us," said one of the leaders. Plant
management was taken by surprise. Plant manager Larry Johnson tried to
convince workers to go back inside and return to their stations.[9]

Peña was holding his usual morning meeting at 9 a.m. when the call came
in from the plant. "The Mexicans are walking out. The Mexicans are walking out!" the caller said. The phone was also ringing at the rectory of St. Andrew Catholic Church, a steepled, red-brick church on the outskirts of Red
Springs, whose priest, Father Carlos Noel Arce, was considered a "guardian
angel" by his Latino congregations in Red Springs, St. Pauls, and Lumberton. Most of his parishioners were undocumented immigrants.

Arce had grown up in Nicaragua, where his father and brother served in
the army, while another brother was a guerilla soldier. The family fled, but
he was able to graduate in theology and philosophy from the Universidad
Latina de Costa Rica in Heredia, then study marketing and business administration in Honduras. That morning, when he answered the phone, an
excited parishioner told him it was urgent. Latino workers at the Tar Heel
plant were going to walk off the job. He needed to come right away. By the
time he got there, between 300 and 350 workers—mostly Latino—were
already outside.

Despite waves of severe thunderstorms, the walkout continued all day,
ebbing and flowing with shift changes and the weather, dwindling to about
two hundred people by 1 p.m., then swelling to as many as five hundred

when the second shift arrived.[10] "As people arrived to go to work, they would decide to either go in or support the walkout, or they went back home, worried about retaliation," said Arce.[11] "I recognized many of them as my parishioners. Others I didn't know. I was watching, trying to understand what was happening."

Workers milled about chanting the Spanish word for justice—*justicia*. Father Carlos was easy to find in the crowd: he always wore his collar and a black suit. He began to realize that the workers, extremely angry about immigration enforcement, intended to shut down the plant. He saw plant officials outside and asked them what they planned to do. They didn't know, but one thing they were certain of, they would not negotiate with the union. "I told them, these people are not members of the union. They are members of my parish." The officials asked if he would help with communication. Arce quickly got permission from the bishop, Vicar Monsignor Michael P. Shugrue, of the Catholic Diocese of Raleigh. The archdiocese offered attorneys to help.

Throughout the day, workers set up a sound system, with microphones and music. They gave speeches, led chants, and carried makeshift signs: "No More Abuse!" "We Want Justice!" Company spokesperson Dennis Pittman, with the help of a translator, tried to talk but was often drowned out by workers. Pittman tried to explain that the company had no choice: it was either cooperate with ICE or be raided, which would have been much worse for the workers. "I think you know what they would have done," Pittman said, as the crowd booed. "I know this is not easy for anybody. I hope you understand we'd be breaking the law if we don't do this." Someone yelled, "The company never followed the law before!"[12]

The scene was chaotic. Workers were still wearing hard hats; ear plugs dangled around their necks. Peña, one of about a dozen union representatives who had infiltrated the crowd, passed out water and pizza. Organizers later joked that he had bought every pizza and case of water in the county. Officially, no hog deliveries were called off, and production slowed but did not stop. Still, hog trucks backed up waiting to unload. "They slaughter 32,000 hogs a day. Once they're out of the trucks, it's hard to stuff them back on," Bruskin said. "They had another 32,000 on the way for the next day and nowhere to put them. It was a huge problem. They were going to have to shut down the plant."

Keith Ludlum led a small contingent of African Americans to join the protest. Ludlum said that he hoped this would be the first of many walkouts. "I think it's fantastic," he said. Dressed in a yellow, blood-spattered water-proof suit that said "Union Time," Ludlum told reporters that workers just wanted the company to be fair. Margarita Vazquez, a Fayetteville resident and worker at the plant, said that she walked out because supervisors were hostile and verbally abusive to Hispanics. "We are not animals; we are people," she said.[13]

Workers were beginning to self-organize. They chose twelve people to represent them. The company said that they would only talk with them through Father Carlos. About 80 percent of his parishioners worked at Smithfield, said the priest. "All my parishioners were outside and people from the union were there and people from the company were there and nobody knew what to do. They asked me if I would serve as a mediator, as a bridge," Arce recalled.[14] "The workers were outside and the company officials were inside, so I went out and in, going and coming with the information." The company refused to meet directly with workers because they said the union was involved. "They would give me something. I went out to tell the workers. They say something else. I came back. It took a lot of time." Each time Father Carlos came back, he communicated what was said to the crowd.

"I used the sound system, which was very powerful, to explain the proposals." Workers called out questions or came to the microphone with suggestions. Gradually, progress was being made. At first, the company said that it was going to suspend any worker who had walked off the job, but the workers refused to leave. "The people decided no. They stayed there the whole night and into the next day. They slept there. Their intention was to stay until they had an agreement."

That night, the union held a meeting with more than fifty workers at the Holiday Inn in Lumberton, where the union organizers stayed. African American and Latino workers shared their demands as Peña translated. Workers wanted the company to withdraw the no-match letters, rehire all fired workers, impose no penalties for striking, and meet with worker-chosen representatives to determine the path forward.

At the end of the day, Father Carlos gave company officials a hand-written list of worker demands, with somewhere between fourteen and sixteen points, he said. Company officials agreed to a powwow—including diocese

For the second day in a row, workers at the Tar Heel slaughterhouse, mostly Hispanic, walked off their jobs to protest the firing of workers whose immigration status was in question. Photo by Raul R. Rubiera, *Fayetteville Observer*, November 17, 2006.

and company attorneys—the next day. "We had built a relationship with Father Carlos," Bruskin said. "He was not a radical, but he was a good guy who cared about his people. He was nervous about the union. Never came out in favor of us. The good news, the company thought he was on their side. So it was to our advantage to allow him to be neutral."

By Friday, the position of the workers had grown stronger as their demands grew more focused. An official negotiation was scheduled for midafternoon inside the plant. Before the meeting, according to one organizer, Father Carlos and his attorneys met with union attorneys in "Roberto's garage" in St. Pauls to narrow down the list of demands.[15] That afternoon, diocese and company attorneys met while the protests continued. Company security, which had been tight the day and night before, relaxed as the crowd ebbed and flowed through shift changes. Union organizers—wearing yellow T-shirts—handed out flyers and got union cards signed. Latino music played over the loudspeakers.

About four hours later, Father Carlos emerged in triumph as between five hundred and a thousand workers and UFCW representatives, disguised as workers in hats and boots, cheered him on. He announced in Spanish that a tentative verbal agreement had been reached. "We agreed that all the workers who had been fired could return for two months," Arce recalled. "The strike was suspended and all the workers were allowed to come back." The tone from Smithfield was conciliatory. "It's a victory for everyone," company spokesman Dennis Pittman said. "They'll be going back to work. They'll be making money. And we'll be operating the plant at full capacity."

Pittman tried to make Latino workers understand that it was hard on the employees in HR and caused them emotional distress to work with ICE and tell good, longtime workers they were going to be arrested. "These are people who have not done anything wrong. Losing them is costing us money," Pittman said. "My whole HR department for the last month has been the most depressing place in the world. The human resource reps are crying; the employees are crying."[16] Production during the walkout had dropped to about 70 percent of capacity, he said. The negotiated agreement gave rehired workers sixty days to resolve identification paper discrepancies. Workers returned for the Friday night shift and production went back to 100 percent. An official agreement was to be signed early the next week.

The next afternoon parishioners crowded into St. Pauls United Methodist Church at the corner of Broad and 4th Streets. The 350-seat church at Red Springs was packed for both morning services as was Father Carlos's afternoon service at St. Francis De Sales in Lumberton.

Father Carlos, wearing his black cassock robes and a green chasuble, told his parishioners what the company had verbally promised him, as a delegate of the Raleigh bishop. They applauded, he said.[17] That Monday, he arrived at the plant with his designated representatives to sign the official agreement. What happened next, he said later, was extremely offensive. The boardroom table was already filled with black and white workers, but there were no Latinos. Dennis Pittman and at least two other supervisors were there. Plant manager Larry Johnson sat at the head of the table. Father Carlos was seated to his right, with his attorney a few seats down. Father Carlos's twelve workers reflected the diversity of those he represented, including whites, blacks, Hispanics, and both men and women. They were told to sit in chairs against the wall. "That was not the agreement. I asked what [those sitting at the table] were doing there and Johnson said he invited them. He said, 'They are workers too.' I said, 'But they are not the chosen ones.'"

Johnson began to read off points from a printed list, announcing which demands the company was meeting, which they were not, and why. The meeting was conducted in English. "I said, 'Wait. Is this a negotiation? Or is this information you invited us to listen to?' He said, 'Well, this is our decision. If you need clarification, we can discuss it one on one, but not with the group.'" Despite Father Carlos's protests, Johnson continued to read, then abruptly ended the meeting. "We were all surprised, and we said, 'Well, what to do?' The twelve said, the most important demands we got. Our attorney said if they wanted to continue fighting they could, but the workers agreed and we left the plant."

Outside, a handful of workers and union organizers, including Eduardo Peña and Gene Bruskin, were there to greet them. Word quickly spread through the plant that the negotiations were officially over and workers had achieved their main goals—that strikers would be hired back without penalty and that fired workers would be hired back and given additional time to address paperwork problems. Pittman later said that Father Carlos was beneficial in helping end the walkout. A news report called him "a bridge between Mexican immigrants and Robeson County authorities" and the "key figure in the Hispanic community."

But Arce has bitter memories from the experience. "There was not a negotiation," he said. "[Plant manager] Larry Johnson was the one speaking. This was a monologue." Arce said that company officials showed "a lack of respect," despite agreeing to the core demands of the workers. "It is not possible to serve the people without knowing their struggles," Arce said of his role. "To be a priest is not just to celebrate the mass, it is to serve and be with the people always."[18]

The victory was short-lived; the momentary win was not sustainable. Smithfield successfully calmed a single storm, while behind the scenes they continued to cooperate with ICE and systematically replaced Latino immigrants with native-born workers. Nevertheless, union organizers believe, the walkout served as a model of collective action that triggered an attitude change inside the plant and spread to newly hired black workers. "That moment clicked in the psyche of the rest of the workers," one organizer recalled. "If these people did this, and the company went ballistic, and nobody got fired, and everyone was rehired, and they got what they wanted, and the company listened. Wait a minute, we can do this too. Looking back. That was the spark that triggered something bigger. It was a transformative moment for the workplace."[19]

Inspired by the company's response to Latino demands, African American workers acted on a long-standing demand of their own — the right to honor an American hero. They collected four thousand signatures from workers and threatened to close the plant if they were not given a paid holiday to recognize Martin Luther King Jr. on January 15. On January 9, forklift driver Leonard Walker took the petition to the plant manager.[20] Management rejected the request and blamed the union for making waves. They criticized Walker for not working through channels. They said they had checked with sixty-two other pork-processing plants and only two planned to close. "The way they responded was that workers could vote for either Easter or Martin Luther King Jr. Day for a holiday," Bruskin said. "Jesus Christ or Martin Luther King, you choose."[21]

When Monday came, hundreds of employees skipped work, many heading to First Baptist Church in Fayetteville in what was called a showdown between the company and the union. Smithfield said that three hundred workers were absent and production was slowed by about 10 percent.[22]

While union organizers said at the time that Smithfield cooperated with ICE to intimidate Latino workers to keep them from supporting the union, that didn't make sense for long-term strategy. It's more likely that Smithfield had little choice. Immigration enforcement was undercutting the company's business plan to use cheap immigrant labor, and black replacements did not come as cheaply. The alternative, like what happened at Swift & Company, was far worse.

———————

On December 12, 2006, ICE raided six Swift & Company meatpacking plants in six states in what was called Operation Wagon Train, arresting 1,300 workers — about 10 percent of the workforce. Workers across the Midwest were bused to detention centers, and most were deported. It was the largest worksite enforcement action in U.S. history. Unlike Smithfield, Swift had been unable to reach a cooperative agreement under IMAGE, despite laying off about four hundred undocumented workers in the nine months leading

Supporters of Smithfield Packing workers marching from First Baptist Church on Martin Luther King Jr. Day to the site where a new statue of King was unveiled. Photo by Marc Hall, *Fayetteville Observer*, January 16, 2007.

up to the raids. Swift had even tried to stop the raids by filing an injunction in November, but it was denied.

The raids were condemned by social justice and religious groups for breaking up families, separating children from their parents, and putting schools in the untenable position of either supervising children after hours or sending them home knowing their parents were missing. Later studies showed that nearly a quarter of Swift's workforce were undocumented and that Swift had to raise wages by about 8 percent after the raids in order to employ legal workers.[23]

Despite the November reprieve at Smithfield, undercover plainclothes officers from ICE raided the Tar Heel plant on January 24 and arrested twenty-one workers. Officers arrived in unmarked cars.[24] Unlike the surprise, public Swift raids, when ICE hit the Tar Heel plant they were discreet, pulling Latino workers off the line individually, saying they were needed in HR. Workers would put down their knives, take off their protective gear, and head to HR, where they were put in handcuffs, locked in temporary detention, then loaded into vans off-site. "They also called up African Americans and whites and told them they had to take drug tests," said Keith Ludlum, who was working that day. "If they'd only called Latinos, people would have known what was happening." If the workers had found out what was happening, hundreds would have dropped their knives and run, leaving pounds of meat to spoil. Word didn't spread until parents didn't show up to collect their kids from school or day care, or didn't come home that night.[25]

"We immediately went into high gear," Bruskin said.[26] "Father Carlos set up hotlines for the families and tried to find out where they had put the people. It took days before we were able to talk to them." Latinos began to panic. Many simply packed up and left, organizers recalled. The *Winston-Salem Journal* called it "an open secret" that the plant had hired so many undocumented workers but said that everyone "was willing to look the other way" because of the economic benefit.[27]

According to ICE, Smithfield—unlike Swift—was not a targeted employer whose business model relied on illegal labor; and Smithfield had been cooperative and provided employee information. Meanwhile, Smithfield announced that another five hundred workers, identified through IMAGE, were to be fired. "There are a lot of question marks about what happened today," Peña told local reporters. "I can't even imagine the psychological effect on the work force. . . . But putting employees who've been loyal to the company for years in such a situation is 'terrorizing' the immigrant community."

On the following day, the twenty-one workers arrested, including one woman, were moved to Stewart Detention Center in Lumpkin, Georgia, nearly seven hundred miles from Tar Heel. "They didn't have lawyers. They didn't know their rights. Most got deported," Bruskin recalled. Production was substantially reduced as workers stayed away, afraid of being arrested. Union workers joined with religious and social workers at the union's worker center in Red Springs to provide help to workers and their families. Making things difficult, ICE refused to release the names of those arrested because they were being administratively charged, not criminally.

When word spread that "La Migra"—Latino slang for the immigration service—might still be in the plant, most of the cleaning crew didn't show up on Wednesday night. As a result, on Thursday morning the USDA closed down the plant. The kill floor was taped off with yellow plastic barriers.[28] In addition, several hundred people did not show up for work on Thursday, the company reported. "It's been a rough day," Pittman told a reporter. "All we were trying to do today is get the product out the door."[29] The company tried to persuade Hispanic workers to return, running ads on a local Spanish-language radio station, saying immigration officials had left. Hundreds gathered at Father Carlos's church in Red Springs, the priest said, looking for answers.[30] "They felt betrayed by the company. . . . They had been working there for years. They were sold out by the company who gave their information to immigration."

Company officials tried to convince workers they had not handed over undocumented workers. One notice read: "We did NOT give ICE the names of people to arrest and we are NOT responsible for any of this. This is the federal government doing what it has decided to do." ICE put out notices that they were not going to sweep Hispanic neighborhoods, but it didn't help. Panicked workers stayed home and kept their children home, refusing to answer their doors or phones.

Democratic U.S. Representative Mike McIntyre condemned undocumented workers and praised federal law enforcement for protecting "our communities from overwhelming costs which drain them of much needed resources for education, law enforcement and health care."[31] According to William Gheen, spokesman for Americans for Legal Immigration in Raleigh, "Illegal aliens have put themselves and their families in this situation by breaking the law." Gheen had publicly called on ICE to raid Smithfield for months. "Let's get them while we know where they are," he said.[32] The Latino community braced for more raids, or simply ran. "It reminds me of an island where a hurricane is coming," Peña said then. "People are getting ready for disaster."

---

By January 28, 2007, union officials estimated, about two thousand workers had left.[33] "We are talking about half the work force. The whole situation is out of control. Workers are in a terrible situation," said Peña. On February 22, the *Fayetteville Observer* reported that the company had announced that ICE IMAGE deadlines would soon expire for another 250 workers.[34]

Smithfield was scrambling to replace workers. The union was scrambling to protect displaced workers while making public statements that Smithfield Foods was using immigration enforcement to break the union. Mark Lauritsen, head of the meatpacking division for the UFCW, said that the

Department of Homeland Security and the company were "worried about people organizing a union and the government said 'here are the tools to take care of them.'"[35]

That Smithfield had any strategy regarding immigration and the union is questionable. They were, it seems, simply trying to avoid massive raids and shutdowns like those at Swift & Company. Regardless, the loss of so many workers who'd been cultivated by union organizers was devastating to the organizers. The heart of the union campaign seemed broken. "It takes years of convincing, or educating people, to develop this kind of trust and activity," said union leader Peña. "The union has become part of the community and backs up what workers want to do. . . . People went from feeling they had no rights to looking their foremen in the eye. Immigrants in particular were taking bolder actions even than citizens. Now people are concerned about basic survival. The message they've gotten is they're nothing. They can be taken from their families, arrested and deported at any time. They wonder who will take care of their kids if the government comes for them. It's hard to think about workplace injuries if that's the big question on your mind."[36]

Meanwhile, the plight of the families was gaining national attention. In early February, the story of Guadalupe was filmed for *The World*, an international news program produced by the BBC World Service and Public Radio International. The 14-year-old high school freshman said that her godfather had been arrested and she was afraid they were going to take the rest of her family away.

Host Matthew Bell went to Father Carlos's St. Andrew's Catholic Church to interview families. "This is a compelling story because it is happening all over America," Bell said during the taping. "The human side is a story that is not often told." As the *Fayetteville Observer* reported, "Bell moved from person to person as they sat in a semicircle in the church's fellowship hall. One by one, they told their story about how they learned of their loved ones' arrest, their thoughts about the crackdown on illegal immigrants and what they plan to do next."[37] One mother of five said her husband had been taken after ten years in the United States; three of their children are U.S citizens. Another's brother had been taken; the two had been sending money back to their families in Mexico.[38]

On February 6, dozens of protestors gathered outside the plant. Reverend

Nelson Johnson of the Southern Faith, Labor and Community Alliance led the rally. "They are forced to endure arrests, break-up of family, loss of income and are thrown into a state of fear," Nelson said as protestors held up signs.

Two days later, the *Winston-Salem Journal* ran a scathing editorial:

A crackdown on illegal workers at a Bladen County slaughterhouse demonstrates the human and economic cost of the federal government's decade-long failure to secure the nation's borders. . . . By finding illegal workers and deporting them, it is sending a message back to Latin America that illegal employment in the United States is getting more difficult. The hope is that that message will slow the influx of additional illegal immigrants.

This is occurring, however, at a significant cost. . . . Families . . . are now left without the chief breadwinner . . . not knowing the whereabouts of those arrested. . . . If the human cost of the crackdown isn't unsettling enough, there's the economic cost to the company. . . . They are good workers, have been good local residents and are fully trained. Replacing them will not be easy, because there isn't a great deal of interest locally in the very difficult work. There has to be a better way to solve this problem, however, than the random capture of a few dozen illegal immigrants out of the millions already here.[39]

That summer, it got even worse. On August 22, 2007, federal agents raided Latino neighborhoods in the middle of the night, arresting twenty people, mostly from Mexico. Eight more were arrested at the plant. The majority were current or former Smithfield workers. Thirteen were women. In several mobile home parks in Robeson County, ICE operatives knocked on doors between 3 and 4 a.m. — sometimes breaking them open — then headed over to the plant just before 5 p.m.

Volunteers with the Workers Center of Eastern North Carolina in Red Springs and St. Andrews Catholic Church spent the next two days trying to track who had been arrested. A community meeting was scheduled at the church, where updates were given and information handed out on how to

Workers demonstrating at the Smithfield Tar Heel plant on February 6, 2007, in support of Hispanic workers who were fired and black workers who wanted to take Martin Luther King Day off. Photo by Raul R. Rubiera, *Fayetteville Observer*, February 6, 2007.

handle immigration raids. The Raleigh diocese paid for legal fees and helped families.[40]

A national film crew, making the documentary *Food, Inc.*, happened to be in town. Union organizer Peña called them during the raids. "They weren't highlighting immigration," Peña said. "They were documenting the food industry. But I got a call at 3 a.m. that immigration just arrested a mother and had her sign a release to leave her kids with a babysitter, so I called and said, 'You want to know what life is like. Let's go.'"[41] Peña and Ludlum went from trailer park to trailer park telling workers not to open their doors, warning them that ICE was there. Ludlum said that both the UFCW and ICE were driving American-made cars — ICE's cars all had Georgia plates on unmarked vehicles — which confused Latino workers about who was who. Immigration agents were "quite frustrated with us," he recalled. "They were trying to keep the film crew back with their cameras."[42]

At least five ICE teams were out arresting workers. "They brought undercover agents from Miami and Georgia to do the arrests," Peña recalled,

Alberta Palmer (center) shouts chants into a blow horn during a protest at the Smith-field Packing plant. Photo by Andrew Craft, *Fayetteville Observer*, July 19, 2007.

Smithfield employees (from left) C. J. and Lenora Bailey and union organizer and former employee Keith Ludlum show their support during a meeting at St. Andrews Catholic Church in Red Springs. The sign on the left reads (in English), "We are united"; the sign on the right reads, "We are with you." Photo by David Smith, *Fayetteville Observer*, January 30, 2007.

"thinking they were in Lumberton in the middle of nowhere in the middle of the night and no one would know them." The undercover agents didn't want their covers blown, so "when we show up filming and all the agents started running for cover, I thought, 'What's going on? Shouldn't we be the ones who are running?'"[43]

At one trailer park, ICE agents cut off power to the trailers. "They tried to smoke them out with the heat index at 110 degrees in August, with children inside," Ludlum said. "It flabbergasted me that they could use those tactics. The people finally gave up, and they took a pregnant woman and a couple of guys out in handcuffs."[44]

The *Food, Inc.* video, time-stamped 4:45 a.m., shows immigration agents arresting Hispanic men at a trailer park. Initially, the figures are shadowy, with the image of a white man with short brown hair wearing a black T-shirt saying in upper case letters POLICE ICE. A brighter scene shows a green trailer at the end of a long, sandy driveway surrounded by tall trees and ragged grass. Seven officers approach, lining up along the steps to a ramshackle wooden porch railing, some wearing black T-shirts, others wearing black vests over white T-shirts identifying them as LICENSE THEFT BUREAU.[45]

They knock on the door, then the agent holding the storm door appears to use a crowbar type of instrument to break the door open. The men, all wearing ball caps, run inside. They come out with two Latinos, handcuffed. A burly white man, slightly balding and with a goatee, wearing a black T-shirt with an insignia, holds the elbow of a short Latino man wearing a white T-shirt with a large "USA" across the chest. Behind them, another agent, his ball cap turned backwards, guides another Latino man — handcuffs in front attached to a waist restraint — wearing a white T-shirt over ripped jeans, his sneakers untied.[46]

"We want to pay the cheapest price for our food," Peña narrates. "We don't understand that comes at a price. These people have been here working ten or fifteen years processing your bacon, your holiday ham, now they're getting picked up like they're criminals and these companies are making billions of dollars. We want to pay the cheapest price."[47]

Those arrested were charged in federal court with identify theft, or using someone else's Social Security number, which can damage the credit of the person whose number it actually is. While the leads initially came from the

IMAGE list of mismatched Social Security numbers, the real owners of the numbers had made complaints to the Federal Trade Commission. United States attorneys said that at least two victims had losses in excess of $10,000; one nearly lost subsidized housing because the worker's wages had pushed the victim over the income limit.[48]

As the *Fayetteville Observer* reported, the immigrants appeared in federal court in Raleigh wearing short-range radio receivers with earpieces feeding them translations. One woman, who was pregnant, was released and carried handwritten messages on torn scraps of paper towels to the families of those held. "For the love of God, please quit work," one said, according to the paper. Another woman was released, and four others were not charged but were detained pending deportation.[49] Fourteen—more than half of those arrested—entered guilty pleas in federal court to fraud and misuse of visas and other documents. Most received up to six months with a $100 fine. The others, after serving their time—most a year and a day—were deported. ICE officials said they had been reviewing names of Smithfield employees for more than a year.[50]

On July 8, 2008, Father Carlos had been appointed vicar of the Hispanic ministry of the fifty-four-county Diocese of Raleigh because of the growing Hispanic population that then made up half the diocese's membership. "It's easy to be together in the church and talk about heaven, without changing what is happening here on earth," Arce reflected in 2016.

> As priests, we should not only celebrate mass, we should follow what Jesus did and be involved with the people. At Smithfield, they don't care about the worker. It's a business. They care about the money. When I went to the plant, they had many pictures there, of healthy hogs. This is clear, that the animals, the hogs have better conditions of life than the workers. Now, I am working with immigrant farm workers. The situation of workers there at Smithfield and now with the farm workers, I've seen that they are exploited. It's the new slavery. It's illegal slavery.[51]

If Arce was right, an emancipation proclamation handed down from the government would be required, and it was far from clear that the government was all that concerned.

## CHAPTER 15

# Stockholders Salvo; Secret Talks; Stalemate

**B**y the mid-2000s, the national labor law was in such bad shape that even a member of the NLRB board was speaking out. Wilma Liebman, one of the three judges who adopted Judge West's ruling in the Smithfield case, began writing a series of articles concluding that the NLRB was not fulfilling the intention of the NLRA, the National Labor Relations Act passed during the New Deal in the 1930s. Liebman was bipartisan. She had been nominated by President Bill Clinton and confirmed in 1997, then confirmed for two additional terms under President George W. Bush. Later, in January 2009, she was appointed by President Barack Obama as chairman, and she served until August 2011.

Liebman believes that Americans have been divided about labor law since it was first enacted as part of the New Deal and have never fully embraced the importance of unions and collective bargaining.[1] Yet collective bargaining, she argues, significantly helps sustain the middle class and the economy. The intention of the law—to balance bargaining power between labor and capital—has not been carried out by the NLRB. As a result, unions turn away from the board and look at other options, including negotiating with employers directly and seeking worker protection from state and local governments, rather than using the NLRB-controlled secret-ballot election process.[2]

During her latter years on the board, Liebman frequently wrote dissenting opinions criticizing the majority on the board for not upholding labor law. "Something has indeed gone wrong. Somewhere along the way, New Deal optimism has yielded to raw deal cynicism about the law's ability to deliver on its promise," Liebman wrote in a journal article. "The National Labor Relations Act, by virtually all measures, is in decline if not dead."[3]

According to Liebman, the fifty-year golden age that followed the passage of the NLRA—the centerpiece of the second portion of the New Deal, with millions of workers achieving middle-class status through collective bargaining to gain fair wages and benefits—was fading. Perhaps most symbolic of the decline was President Ronald Reagan's firing of striking air-traffic controllers in 1981, which Liebman called "a watershed event." The economy transformed as oil prices climbed, technology exploded, and foreign trade increased.

"What followed over the next two decades is familiar," Liebman wrote. "The Cold War ended; technological innovation accelerated; relentless competition, both domestic and global, grabbed the economy; major industries were deregulated; manufacturing declined and the service sector exploded; shifting demographics changed the composition of the workforce; and a fourth wave of immigrants crossed our borders."[4] The model of manufacturing plants in the 1930s and 1940s became "anachronistic in a post-industrial and fiercely competitive global economy." The social contract was broken. Weakened labor resulted in the near disappearance of the strike as a tool for achieving equality, while businesses became strong and sophisticated in overcoming unions' collective action.

Liebman blamed the NLRB for not keeping pace with a changing workforce and technology. The line between worker and supervisor had become blurred. With the surge of independent contractors, even the notion of protection of employment or work conditions had disappeared. In addition, the inability of the NLRB to enforce its actions and its dependence on the courts, which routinely overturn its decisions, further undermined the intention of the NLRA. "Today, the perceived obsolescence of the Board is linked in substantial part to its seeming lack of administrative will," she wrote, citing the board's inability to award back pay and the delays in the legal process that deny justice. The Seventh Circuit famously referred to the NLRB as the "Rip Van Winkle of administrative agencies," she noted, adding that "the problem here, however, is not the statute, but the agency that administers it . . . [whose] decisions suggest an underlying discomfort with government regulation of business, the notion of collective action, and the zeal that may accompany those efforts: the fundamental premises of this statute."[5]

Liebman had high hopes that the Employee Free Choice Act, which was reintroduced by each Congress through the 111th serving until 2011 (and again unsuccessfully in 2016 in the 114th Congress), would educate Americans about "the erosion of the right to organize and the danger posed to our society as a consequence, especially in the context of growing income inequality." Ultimately, she believed that the NLRB could be salvaged and once again fulfill the promise of the NLRA. "Like dinosaur DNA, the promise of the Act is worth preserving. The stakes are too high to do otherwise."

Spurring on legislators was evidence of a growing income gap in which the richest 1 percent of Americans had captured most of the economic growth in 2005, while the bottom 90 percent had captured none.[6] Further evidence that the NLRB was significantly undermining and underrepresenting worker protection was a significant reduction in the number of unionized workers in the private sector to the lowest point since the NLRA was passed. Organized labor had declined from about 35 percent in the mid-1950s to less than 8 percent by 2007. In addition, the number of cases brought before the board had fallen by 26 percent from 2005 to 2006. From 1997 to 2007, the number had declined by 41 percent. During that same ten-year period, unfair-labor-practice case intake declined by 31 percent.[7]

Wilma Liebman and Keith Ludlum, though worlds apart in terms of education and profession, were connected not only by the case but by their opinions. With one looking down from the government bureaucracy while the other looked up from the workplace, Liebman and Ludlum had reached identical conclusions. The NLRB was broken, and the Employee Free Choice Act would go a long way toward returning labor law to its roots. Though the main point of the act was to force companies to recognize a union without an election, it also included stiff penalties for illegal actions by employers, up to $20,000 per violation. Additionally, it set a time limit of ninety days on negotiations.

One of the best points of the proposed law, Ludlum thought, was that it would require that companies be fined thousands of dollars for every day if a worker was found to have been fired illegally, a far cry from the percentage of lost wages after subtracting comparable pay that workers receive after firing under current labor law. The average amount companies currently have to pay illegally fired workers is about $3,500 to $4,000.[8] "They'd have to think real hard about firing me. $365,000 a year for twelve years would make an employer hesitate," Ludlum said.[9]

The chairman of the subcommittee, Robert E. Andrews, a Democrat from New Jersey who came to office in 1990 and co-sponsored the bill, cited the shrinking middle class together with the fact that union workers make about

30 percent more and are much more likely to have health insurance and a pension than non-union workers. According to the AFL-CIO, union workers are 63 percent more likely to have health insurance, four times as likely to have a pension, and seventy-seven times more likely to have short-term disability benefits.[10] Andrews underscored what Gene Bruskin had been saying all along: secret-ballot elections are a joke. In fact, corporations opt for secret-ballot elections so they can illegally coerce workers, undermining the original intentions of labor law.

---

On February 8, 2007, Ludlum testified before the House Committee on Education and the Workforce's Subcommittee on Health, Employment, Labor and Pensions about the latest version of the Employee Free Choice Act, once again pending before the House.[11] The union staff helped Ludlum prepare his statement and prepare for the session, but nothing could have prepared him for the formality and the pomp and circumstance of the proceedings. The ceiling soared above the rail-and-stile paneling and large portraits lining the walls of the Rayburn House Office Building in the Capitol Complex. Democrats sat on the right, Republicans on the left, with the chairman in the center, sometimes rudely firing questions and demanding instant responses. "My hands were shaking whenever I went for the water," said Ludlum. "It was nerve-racking when you don't know what's going to happen."[12]

There was some humor. Before he began, Ludlum was instructed that a bank of lights would let him know how much more time he had to speak, including a yellow warning light and a red light indicating it was time to wrap up. "If the red light goes on for too long, a trap door exists beneath your seat through which you will fall." The audience laughed and Ludlum relaxed a bit.

"My service in Desert Storm was to protect the laws of our land and not to protect companies like Smithfield that continually violate those laws," Ludlum testified, telling of his illegal firing twelve years before and Smithfield's use of violence in 1997. "If anybody thinks that this company is going to have a free and fair election after its history of violence and intimidation, then you haven't heard a single word that I have said."

Ludlum pointed out that Smithfield—or any of its executives—had never been punished, fined, or indicted for breaking the law. "Smithfield was only required to offer jobs to those workers like me who were illegally fired and pay back-wages for the time we were unemployed or could not find comparable pay." He went on: "A $1.5 million settlement . . . for a company that sells $11 billion worth of products a year, this is pennies in a bucket. . . . There is simply nothing to deter them. . . . The company has been fined by OSHA. It has been fined by the Environmental Protection Agency. It has been found liable for massive violations of labor law. But breaking the law is just the cost of doing business for Smithfield. . . . At Smithfield, the USDA is present in the livestock yard to ensure that the hogs are not abused or unduly stressed," Ludlum testified. "I witnessed Smithfield repeatedly putting more value on the hogs . . . than the workers' health, safety, and well-being."

Others who testified at the hearing vehemently disagreed with the proposed bill. They argued that moving from a secret ballot to an open-identity check card made workers vulnerable to union intimidation, since the union would have a list of who hadn't signed on. Virginia Foxx, a Republican representing North Carolina's Fifth District, asked Ludlum if Smithfield had agreed to hold a fair and open election by secret ballot. Ludlum responded: "They definitely want a fair shot at keeping us in the secret-ballot process because it was twelve years from the first one, ten years from the second one and they can postpone, you know, a legitimate election for another twelve, fourteen years, you know, have twenty, thirty years operating and abusing workers. . . . Sending us back for another NLRB election is sending us back into the lion's den." One member asked Ludlum, "When it takes as long as a decade to finally get reinstated, what is the message that the law sends to employers and employees about the value of unions?" Ludlum replied, "I am a shining example for Smithfield to the other workers. If you speak up, you stand up for your rights, we will fire you and we will see in twelve, thirteen years." Those against the check-card system argued that high-end officials in the union could use aggressive tactics to pressure top-level executives to make deals based on economic pressure or corporate campaigns that were not percolating up from the workers.

Most of the testimony and question-and-answer period was spent bemoaning the broken NLRB and the loss of the middle class. Since 1995, 6 million Americans had lost health insurance; meanwhile, corporate CEOs earned 262 times as much as average workers—35 times more than in 1978. In 1969, six thousand workers were retaliated against illegally for supporting a union;

by 1990, that number had grown to 20,000. Anti-union consultants are hired by employers in 75 to 82 percent of worker campaigns to form unions.[13]

After two and a half hours, Ludlum was feeling both exhausted and exhilarated. He had rubbed shoulders with powerful people and stood up to intense questioning, but he was also disappointed. He thought that the best points of the card-check system had gotten lost in the long-winded attacks on the NLRB as a whole. He drove back to his double-wide trailer in Bladenboro and was back driving hogs by 5:50 a.m. the next morning, getting splattered with hog shit and stirring up trouble.[14]

On February 13, the Committee on Education and Labor voted 26 to 19 to send the bill to the full House. Republican members voted unanimously against the bill. On March 1, 2007, the bill (HR 1409) passed the House by 241 to 185. Ludlum was ecstatic. On March 30, Senator Ted Kennedy (D-MA), chairman of the Senate Committee on Health, Employment, Labor and Pensions introduced the Senate version. Senator Barack Obama urged his colleagues to pass the bill:

> I support this bill because in order to restore a sense of shared prosperity and security, we need to help working Americans exercise their right to organize under a fair and free process and bargain for their fair share of the wealth our country creates.
>
> The current process for organizing a workplace denies too many workers the ability to do so. The Employee Free Choice Act offers to make binding an alternative process under which a majority of employees can sign up to join a union. Currently, employers can choose to accept—but are not bound by law to accept—the signed decision of a majority of workers. That choice should be left up to workers and workers alone.[15]

---

On June 26, despite high hopes, Democrats were unable to pass the bill. Republicans successfully turned the argument on its head, saying that it stripped workers of their right to privacy and their right to vote secretly without pressure from unions. Ludlum was extremely disappointed and funneled his frustration into improving work conditions for his immediate co-workers. He lobbied for a break room and clean drinking water for Livestock. He circulated a petition to OSHA, then organized a one-day work stoppage in June. "People were eating in the locker room and bathroom, having lunch where people had dried hog shit on their clothes. Drinking out of coolers that had hog shit on them. People with hog shit on them were making the water coolers. I wasn't drinking their water and I wouldn't eat at work," he said.[16]

Ludlum was asking a lot of questions. What protocol was in place to sanitize the coolers? What solution was used to sanitize the coolers? Ludlum told his colleagues that there were two kinds of mad: mad enough to talk about it and mad enough to do something about it. Furthermore, concerted activity regarding safety issues is allowed under federal labor law. Ludlum assured them that they couldn't be fired.

They held a vote and elected Ludlum and Terry Slaughter to be the spokespeople. They agreed that when they got to work the next day, they would clock in, then go to the grass area and refuse to run the hogs. "If there ain't no hogs going into the plant, then the plant is not running. I'm in the most critical department. I've got them like this [gestures with his right hand under his testicles]," said Ludlum later. "I can send a message to the whole plant."[17]

With twenty to thirty workers standing down, Ludlum demanded to talk to the plant manager, a vice president with Smithfield Foods. It took about an hour, he recalls, for the plant manager, the HR manager, and the safety manager to walk out onto the grassy area where the workers waited. The managers tried to talk to various individuals, but Ludlum had told them ahead of time, "They'll be looking to snag the weak ones. Don't be shaky about it. Nobody move. Nobody say nothing." According to Ludlum, after shutting the plant down for a good two hours, management was out there running hogs themselves in slacks and ties. "Supervisors saw, Livestock managers saw. We had tried to get meetings and they wouldn't meet. Now they were there on our terms."[18] As a result, Livestock got new plumbing, new locker rooms, and a new system for drinking water.[19]

Ludlum believed the company had planted employees in his unit to report on his activities. That August, one of them walked into the new break room and the other guys started calling him a snitch. Ludlum recalled about twenty or thirty workers—all African American—razzing him. The man headed straight for Ludlum. "Don't be calling me that," the man said. "Stop snitching and they'll stop," Ludlum retorted. The man stepped toward Ludlum and pushed him. "Don't be pushing me," Ludlum said. The man pushed him again. "You need to stop pushing me," Ludlum said.

The man pushed him a third time, then lifted his fists up into boxing position. Ludlum lost control. "I can't take a punch," Ludlum recalled. "It was on. I cut him over the eye, broke his nose. He was bleeding out of his mouth. I gave him three chances. Everyone told management Keith didn't have no choice." He continued: "I got years of defensive training in the military. He came at me like a boxer. The first thing I did was kick him. He bled like a stuck pig. It took two hours to clean up the blood in that break room."[20]

Both Ludlum and the worker were fired. Ludlum thinks the man who incited him was made promises or paid off or didn't realize he would have to be fired for fighting too. "He didn't realize they have a workplace violence rule. It doesn't matter who started it." There was nothing the union could do to help Ludlum — except hire him as an organizer, get him the hell out of town, and let everybody cool down.[21]

Despite the failure of the Employee Free Choice Act to be passed by Congress, Bruskin believed the union could create enough pressure to force Smithfield to agree to a neutral playing field in a secret-ballot election that would include equal access to workers and immediate intervention by the NLRB if unfair labor practices were charged by either side. The next phase of the external campaign was to put pressure directly on stockholders. It was perhaps the trickiest and most delicate undertaking so far.

For years now, Bruskin's team had been sending stockholders information about the company's labor policies that could potentially impact the bottom line, and politely raising questions at annual shareholder meetings, where investors discussed profits and problems within the company. But in August of 2006, Bruskin's team ramped up, publicizing a protest on radio, through social media, and in newspapers. As protestors gathered outside the event at the Jefferson Hotel in Richmond, Bruskin announced, "We've come here to send a message to Smithfield Foods while their board of directors and top executives gather to talk about their success and growth of the multibillion-dollar company. We want to remind them that there are people suffering every day in the largest meatpacking plant in the world."[22] Smithfield said that the union purposely obfuscated who was behind rallies and framed protests as being worker driven, rather than union driven.

On September 18, 2006, organizers also mailed a letter to financial analysts who provided coverage on publicly traded companies in the food industries and maintained a purchase rating on Smithfield Foods stock.[23] The letter was signed by Bruskin on Justice@Smithfield letterhead, with a return address of the union center in Red Springs near the Tar Heel plant. Bruskin's team enclosed newspaper clippings and information about the workers' desire to organize a union.

"The campaign's position was that Smithfield's emphasis on defeating the workers' right to unionize was a terrible business position," Bruskin said. "It was bad for the brand and was draining huge amounts of financial resources and interfering with production in their mother plant at Tar Heel. We didn't feel like they were telling investors and analysts the full story, so we sent periodic mailings to the investors that used public records, public articles, and reports on the campaign and other related information. We didn't advocate. We didn't say pull your stock. It was just FYI."

The letter described employee relations as "characterized by unlawful assaults, firings, racial manipulation, intimidation and threatening arrest by Federal immigration authorities by Smithfield." Smithfield said that the letter's obvious purpose was to "negatively influence its recipients' analysis and coverage of Smithfield Foods stock and thereby to injure Smithfield by causing a reduction in the value of Smithfield stock."[24] The letters continued for several years, but Bruskin said he doubted that they ever made a difference in stock values. "I don't think analysts specifically downgraded them because of us," Bruskin said. "The price of hogs and grain was huge and had a bigger impact."

Still, the cumulative effect of the corporate campaign had finally brought Smithfield to the table. As Bruskin planned a large protest at the stockholders meeting in August of 2007, secret high-level negotiations were underway between officials of the UFCW, including UFCW executive vice president Pat O'Neill, and Smithfield Foods executives. Some negotiations were face-to-face, supported by phone calls and emails. "The fact that they wanted to engage with us was encouraging," Bruskin recalled. "But there were a lot of up and back proposals with no substantive movement." Some tentative agreements were put in writing, but nothing was legally binding and there was "way too much vagueness," Bruskin said.[25]

Now that Joe Luter was no longer in charge, after stepping down in 2006 immediately after the U.S. Court of Appeals upheld the NLRB's finding that Smithfield's fight against the union had been illegal, the union believed that Smithfield would be more receptive to agreement. The union pressed for their bottom line: rules that would create a nonthreatening environment for a secret-ballot vote, accompanied by agreements for immediate rulings by the NLRB rather than lengthy appeals. But, Bruskin recalled, the union was "pushing all the details and (Smithfield was) just stalling."[26] The union wanted the secret-ballot vote to be at a neutral site. It was a sticking point: Smithfield insisted the election had to be at the plant. Neither side would budge.

Meanwhile, during secret negotiations, union higher-ups had been putting pressure on Bruskin to cancel the planned protests at the stockholders meeting that August. Bruskin argued and won. "My position was these are unique moments, once a year. We have to keep doing what we are doing."

---

On August 29, 2007, the union led a "caravan" of labor activists—hundreds of union members, students, clergy, and plant workers—to protest at the annual shareholders meeting in Williamsburg.[27] During a rally before the event at the First Baptist Church in Williamsburg, the crowd overflowed through open doors into the churchyard. Most wore the bright-yellow signature T-shirt with the Justice@Smithfield logo.

Reverend Nelson Johnson, who had organized clergy and rallied support for the workers for so many years, raising awareness and standing for hours protesting outside the plant, spoke first. "We have heard of the dangerous conditions on the kill floor. . . . We have heard of your inability to get clear drinking water on a sweltering hot day. . . . We have heard of the midnight raids with officers surrounding the homes of our beloved sisters and brothers, separating them from their children. . . . This business of immigration is a shameful national outrage."

Some jumped to their feet as applause broke out.[28] Johnson continued preaching: "We are going to take this message on the authority of your suffering into the boardroom of Smithfield and declare on this day we are calling on you to change your ways and to conform to the righteousness of God."[29]

Next was Reverend Dr. William Barber II, the activist black minister and national board member of the NAACP who later spoke at the Democratic National Convention in 2016. Barber preached in traditional black church–style call-and-response, evoking a spiritual revival more than a demonstration, with shouts of "Amen!" and standing ovations by more than fifty Smithfield workers.

"We will fight with weapons of our own warfare," Barber preached. "Which is truth and justice. The NAACP is not for sale. Our dignity is not for sale. We don't want pork. We don't want pork money. We just want people to be treated right." According to the *Fayetteville Observer*, "the crowd roared" when Barber told them of a $1,000 check sent by Smithfield Packing to the Fayetteville chapter of the NAACP, which returned the money.[30]

"Smithfield, we need to move forward, stop the propaganda and start the process with the union and the workers," Barber preached. "Put away union busting consultants and come to the table with union workers. Stop dividing black and Hispanic and develop a plan to pull all of us together."[31] "Twelve years is long enough. Twelve years of denial. Twelve years of disruption. Twelve years of delay. Twelve years of deliberate actions against our workers." Barber's voice boomed out. "We are not confused. We need to move forward. Break the chains of injustice. Break them!" The preacher's voice rose. He shook his head back and forth, beginning to shout in a raspy, urgent voice. "When Smithfield does this, we'll stop marching! When Smithfield does this, we'll stop protesting! When Smithfield does this, we'll stop organizing!" Barber leaned into the microphone. "But until then!"[32] Protestors stood and began to sing "We shall not, we shall not be moved," swaying with syncopated clapping.

The protest moved from the church into the streets outside the shareholders meeting at the Williamsburg Lodge, a resort hotel on South England Street, with eight hundred to one thousand workers and activists chanting, "Down with Smithfield! Up with Justice!" As uniformed patrols and mounted police observed, workers—many wearing T-shirts, shorts, and tennis shoes—blew whistles, banged on drums, and held up signs in the shape of pigs that read, "Break open Smithfield's piggy bank." In one video, a man held up a sign that said, "Workers' rights are human rights."[33]

By now, the UFCW campaign had the full support of not only the North Carolina NAACP but the national organization as well. Barber, on the national board of the NAACP, helped lead the protestors. When Smithfield attempted to prevent union stockholders from entering the meeting, despite their legitimate stockholder proxy passes to the event, Barber began to pray in his deep resonant voice in the hallway outside the meeting room. His booming voice called out to God to influence the outcome of this impasse.[34] "They changed their mind and let us in," Bruskin recalled.[35] Once inside, about a dozen activists and Smithfield workers presented a signed petition demanding a fair and neutral union vote and contract. "We got three thousand signatures demanding the company create a level playing field and allow workers to make their decision without being harassed," Bruskin said later.[36] The petition was also signed by Danny Glover, Susan Sarandon, Al Sharpton, and Jesse Jackson.

Inside, Ludlum and his colleague Terry Slaughter were amused and surprised by the group of about 150 stockholders, many of whom gathered around the windows looking out at the hundreds of protestors in yellow Justice@ Smithfield T-shirts. "It was real funny to see all this variety of people," said Ludlum. "You know they're all rich, but some are trust babies, and some are—you could actually tell by their expressions—empathetic to the workers' plight."[37] Smithfield executive Larry Pope referred to the ongoing—now not so secret—negotiations between the union and the company, saying officials were optimistic of nearing a resolution. "We are tired too," he told the activists. Both sides had agreed that employees needed a union chapter, reported the *Fayetteville Observer*, but were "at a standstill on how to create it."[38]

Worker and union activist Terry Slaughter publicly asked Joe Luter: "I have ten co-workers in here and three thousand outside. They want to know why won't you let the workers in Tar Heel have a union?"[39] Ludlum recalls Luter being rude and talking about unions threatening people on bridges, "like Mafia stuff." Luter, who by now was the nonexecutive chairman of the board of directors, dismissed the idea of a card-check system. "Someone comes to your house at nine at night and says, 'Sign this,' a lot of people are going to sign it just to get the person off the front porch," he said, calling a secret-ballot vote "as American as apple pie."[40] Reverend Nelson Johnson found the atmosphere appalling. "It was very tense. I've never been treated so rude when I stood to speak. There was an armed guard next to us."[41]

———————————

When high-level negotiations resumed the next day in Washington, "needless to say, they were very very angry that we did that at the shareholder meeting," Bruskin said.[42] Public statements regarding progress on the talks were cool and restrained on both sides. The sticking point was still how to allow workers to decide: by secret ballot or by card check. The company said that a card-check system would not be private and would be susceptible to union intimidation. The union said that a secret ballot would allow for company intimidation.

"During negotiations, there were lots of discussions inside the UFCW about what kind of conversations we could have and should have with the company," Bruskin recalled. "This is a company we had relationships with at their other plants. It's not like we don't know them or who the players are. We just weren't willing to go to the table and beg. We didn't want to send that signal."

By October 16, six weeks of talks officially ended in a stalemate. Both sides blamed the other. The company said that it had offered a multipoint proposal to the union on September 23. The proposal included many concessions, but the company held fast to a secret-ballot approach. The union in turn held fast to their position that a traditional NLRB election would not protect workers from intimidation. Without concessions on the issue of neutrality, they refused to move forward.[43]

"That's a song and dance that has worked for them since 1994, when they had the first election. And in which they were found to be violating the law, and they said, let us have another election. We promise—they put it in writing. We promise to obey the law," Bruskin said. "And that was the second election. And where they totally violated the laws, they were off the chart. They then appealed that for a decade. And now they're turning around like the man that beat his wife for ten years and says, 'I'll stop now. Do you trust me?' And the workers' answer is, no way."[44]

But Smithfield had been working secretly for months on another way to defeat the international union and possibly deliver a near death blow. Bruskin now believes it's possible that the negotiations were a stalling technique and were never in good faith.[45]

## CHAPTER 16

# RICO Shocks; Both Sides Flinch

O n October 17, 2007, Smithfield filed a multimillion-dollar private civil RICO (Racketeer Influenced and Corrupt Organizations) suit against the UFCW international union, the UFCW Local 400, the Change to Win federation, the Research Associates of America, Jobs with Justice, and seven individuals—including Gene Bruskin.

The union didn't see it coming. First came shock, then anger, Bruskin recalled. "It turned out, they had been working on this for months," he said. "It's possible if we had signed on to what they wanted, they wouldn't have filed suit. But it came as a real kick in the teeth, because we were trying to negotiate honestly."[1]

The suit claimed that the Justice@Smithfield campaign Bruskin had directed had damaged the company's brand name and devalued company stock. Smithfield's suit asked for a minimum of $5,900,000 in damages, plus three times the cost of litigation, later estimated at $25 million.[2] Smithfield alleged that the defendants were guilty of extortion under North Carolina and Virginia law.[3] The suit outlined every element of the UFCW campaign including the use of the RAA publication "Packaged with Abuse," which it claimed was a false report about safety and health conditions. It noted the campaign against Harris Teeter, the national boycott, the resolutions by major cities, the protests against Paula Deen, protests at stockholder meetings, and letters mailed to financial analysts.

From the suit, it seemed as if Gene Bruskin and his Justice@Smithfield team were solely responsible for "injuring Smithfield economically until Smithfield either agreed to Defendants' demands or was run out of business." "Fully knowledgeable of Smithfield's successful business relationship with Harris Teeter, Defendants specifically focused on destroying this relationship as part of their extortionate scheme," the RICO suit said.[4] The Harris

Teeter Father's Day event, according to the company's suit, was part of a pattern of extortion. An article posted on the Justice@Smithfield website "misrepresented the nature of the event as originating with the 'children of Smithfield workers,' when in fact the event was created and orchestrated by the Defendants."

The suit named twenty-one people, including Eduardo Peña, who managed much of the ground campaign at Tar Heel, and alleged that the scheme had begun in September of 2005 with the formation of the Justice@Smithfield campaign and continued with the "Corporate Campaign" in June of 2006. Corporate campaigns, in the words of the suit, "involve the investment of substantial resources towards events, actions and conduct that have little to no direct relation to the wages, hours, benefits and working conditions of the employees of the target employer."

The suit accused the defendants of creating sham entities and portraying their actions as spontaneous, recruiting disgruntled employees and portraying them as representative of all workers, and exploiting the genuine social political concerns of third parties including civil rights organizations, environmentalists, and consumers. It said that Bruskin carried out the scheme for the conspiracy.[5]

Within twenty-four hours, Bruskin was meeting in Washington with union leadership and lawyers. Pat O'Neill, executive vice president and director of organizing, was there, along with other senior executives in the conference room of the legal department of the UFCW. "My first reaction, aside from being pissed at the company, I was nervous," Bruskin said. He thought UFCW officials might end his campaign, and while they warned him to be doubly careful, they gave permission to proceed. "Aside from contracting with a top-notch law firm to assist our attorney, we agreed we had to be doubly careful of everything we said. We recognized this was going to be a potentially huge amount of work and cost. Just to fight the legal battle, they were going to bring in major legal power."[6]

The union did not believe RICO would destroy them, but the financial penalties if they lost would be devastating to both the UFCW and the labor movement as a whole. Unions have long relied on the First Amendment guarantee of freedom of speech. If that tool were determined to be extortion, it would open up a new era of corporate power against the unions.

The RICO lawsuit was a creative effort to kill the union campaign. It had been used at least six times before, according to G. Robert Blakey, one of the attorneys who coordinated the RICO suit for Smithfield and had helped draft the original RICO legislation as a staff lawyer in the Senate.[7] Judges had permitted five of those six cases to proceed, and the parties had settled. "When they settle," Blakey said, "it normally breaks the campaign."[8] Without words such as "falsely" or "sham," the activities cited were traditional labor-union activities. Also, there was none of the violence or threats of violence normally associated with RICO, at least not on the part of the defendants.[9]

News reactions to the suit were dubious. The *Daily Press* in Newport News, Virginia—in a news report, not an opinion piece—said, "Smithfield started this fight by using bare-knuckle approaches in two union campaigns . . . [and] was rebuked heavily for its behavior, but now it portrays itself as the victim."[10] The *New York Times* quoted a New York City councilwoman who sponsored a resolution against Smithfield saying that the practice Smithfield called racketeering was lobbying, or the First Amendment right to petition the government.[11] "Smithfield's 94-page lawsuit sputters with an outrage not always grounded in a sure command of the English language," the *Times* reporter wrote, citing the suit's characterization of a union representative's argument about a water permit as "salacious," a word that generally means sexually stimulating or obscene.

The *Times* also poked fun at the suit's example number 220, which alleged that the union published a defaming article. The alleged defamation was included in a quote that said Paula Deen's "approach to learning that her sugar daddy regularly abuses its workers is to clap her hands over her ears and say, 'I can't hear you!'" It also mocked the suit for alleging that using a quotation from Upton Sinclair's *The Jungle*—"It is difficult to get a man to understand something when his salary depends upon his not understanding it"—was a defamation of Paula Deen. "That's right: Smithfield maintains that it is a form of racketeering to quote an American Master."

The paper interviewed Blakey, who wrote the suit for Smithfield. "It's economic warfare," Blakey said. "It's actually the same thing as what John Gotti used to do. What the union is saying in effect to Smithfield is, 'You've got to partner up with us to run your company.'"[12] Bruskin was quoted in response as saying, "If we kidnapped the C.E.O. and we said 'We know where your children go to school,' that's a Mafia-like act. . . . If we told the truth about how the company abuses workers to its customers, that's traditional free speech.'"[13]

The suit was placed on the rocket docket in the United States District Court for the Eastern District of Virginia in Richmond, one of the country's most conservative courts. Judge Robert E. Payne reviewed a series of motions leading up to the October 2008 trial date. Payne, who received both his BA and JD from Washington and Lee University, was nominated by President George H. W. Bush and confirmed in 1992.

In the defendant's motion to dismiss, the union pointed out that in 2006 the United States Court of Appeals for the District of Columbia (DC Circuit) had upheld the full NLRB board ruling from 2004 (which upheld nearly all of the 2000 decision by Judge West). The defendants pointed out that the federal appeals court had additionally found that Smithfield "had engaged in such egregious or widespread misconduct as to demonstrate a general disregard for the employees' fundamental statutory rights." Their broad point? The campaign had all been based on the facts, as ruled on by the NLRB and federal courts. Citing those facts was protected free speech. Further, they argued that a moral victory could not be extortion. Smithfield argued that the union had substantial financial stakes. Specifically, they argued that an employer's right to recognize or not recognize a union is property and the "right . . . to make a business decision free from outside pressure wrongfully imposed" was an "intangible" right.

The crux of the legal decision in front of Judge Payne was the question of what property was at the center of the extortion. That, he determined, was Smithfield's ownership of the "property right not to recognize the UFCW" and that the defendants had sought to deprive them of this right through extortion.[14] On these grounds, on May 29, 2008, Judge Payne rejected the union's motion to dismiss the case, agreeing that the right not to recognize a union was tangible property and that the conspiracy aspect of the RICO statute had been met. "The alleged predicate acts had the same or similar purpose," he wrote in his decision. "The acts were participated in generally by the same people. The alleged victim was the same. The alleged method of commission was the same or similar," thus fulfilling the conspiracy requirements of RICO.[15]

While it may seem strange to determine that something as intangible as the right to recognize a union could be considered property, the "right to recognize (or not) a union as bargaining representative is among the most valuable and important of rights by business owners," said labor scholar Julius Getman.[16] But, Getman argued, the NLRA had taken that property right away from employers and given it to employees. It is the employees who determine whether they want union representation, not the employer. If Payne continued down this road, Getman argued, it would be the unions who could use the RICO statute against employers who seek to take away that right through intimidation, violence, surveillance, firings, threats of closure, and other illegal labor acts.

Critics decried the use of RICO—typically a tool against organized crime—as a substitute for labor relations law under the NLRA. "If followed, Judge Payne's opinion in the Smithfield case would turn RICO into a labor-relations statute. . . . [This] reasoning would give unions a strong basis for arguing that serious employer unfair labor practices aimed at overturning a union's majority status also violate RICO," said Getman.[17] "Another disturbing feature of Judge Payne's opinion is the importance that it gives to allegations of false and misleading statements. In the absence of such allegations, it would have been apparent that much of what the union did was typical First Amendment protected speech and protest."

---

Meanwhile, in Washington, pretrial activities were in full swing. Defendants were inundated with giving and reviewing depositions, and reviewing and handing over thousands of pages of discovery materials. All UFCW staff were called, along with all upper-level plant managers. The depositions were "massive events," Bruskin said, describing the surreal atmosphere of being in "a giant room, with my own personal lawyer, the union lawyers, [and] about eight company lawyers." There were stenographers transcribing every word, which appeared almost instantly on laptops around the room. The company's attorney had a pile of Bruskin's emails stacked about two feet high, looking for brash statements—undoubtedly plentiful—that could have been part of an extortionist plan. Every public statement may have been carefully vetted, but personal emails among revolutionaries were undoubtedly loaded with vitriol.[18]

Discovery ended in August 2008, and lawyers on both sides began final preparations for the trial, scheduled to begin that October. Some said that Payne just wanted the case to go away. "There is reason to believe that Judge Payne was uncomfortable with the publicity that the case received. He strongly encouraged the parties to settle the dispute."[19]

That September, negotiations for settlement continued. Both sides had a lot to win—and a lot to lose. "The company offered to settle if the union would agree to pay their legal costs of $25 million," Bruskin said, estimating Smithfield's final costs for the RICO suit at $35 million. "We weren't going to pay them for suing us."[20]

Still, Smithfield had a lot to lose by going public. Research Associates of America, the former affiliate of the AFL-CIO, had OSHA logs, testimony of workers, medical records, workers' compensation records, and a myriad of documents to back up their negative campaign. Bruskin believed that a long, drawn-out trial with a lineup of injured workers describing horrific conditions in the plant was something Smithfield was desperate to avoid. "The company realized that all that was going to come out in court." Additionally, Smithfield was beginning to realize that Judge Payne was not going to continue ruling in their favor. "The judge didn't want to get overruled and look foolish," said Bruskin.[21]

Motions prior to trial included allegations that Smithfield had hired an agent to infiltrate the union's campaign and pass along confidential information on strategy. In a key ruling, on October 23 Judge Payne held that to collect its claim of $900 million in damages, Smithfield would have to prove that statements made by the union were false and that they acted with malice, the same conditions as for a libel suit.[22] As Bruskin said later, "There were indications that even with this judge the issues of free speech and defamation would be central."[23]

While Smithfield prepared for court, something much larger than their arguments in court loomed over them, portending defeat. Smithfield had to face the distinct possibility that a pro-union African American candidate, Barack Obama, who had already aligned himself with the workers at Smithfield, was going to become president of the United States. On May 2, 2008, in a joint appearance in Raleigh with Hillary Clinton, Obama praised members of the union, specifically naming those trying to unionize the Smithfield plant in Tar Heel. In what appears to be a pretaped video, Obama looks at the camera and says, "UFCW, you and I share a vision for America. We believe that every American should have a fair shot at life."[24] Similarly, on April 17 in Greenville, Obama told the *Fayetteville Observer*—specifically responding to questions about the Tar Heel plant—that he favored a system that bypassed secret ballot and allowed workers to unionize if a majority signed up. "Obama said companies have learned how to delay, stall and intimidate union efforts. . . . The Bush administration has not enforced legislation affecting unions. It will be a different philosophy when I'm president."[25]

The trial was scheduled to start on October 27, just eight days before the presidential election. "The handwriting was on the wall," Bruskin said. "Looking at a Democratic Congress and president viewing Smithfield as the paradigm of abuse was bad business."[26]

The night of October 26, union staff and other defendants arrived at the Commonwealth Park Suites Hotel across the street from the courthouse. In the early hours of the morning, a handful of principals and their lawyers hammered out an agreement. President of the UFCW Joe Hansen and Pat O'Neill, executive vice president, were present, Bruskin said, along with Timothy O. Schellpeper, president and COO of Smithfield Packing, and George H. Richter, president and CEO of Smithfield's Pork Group.

The two sides agreed to an NLRB secret-ballot election. During the vote, Smithfield agreed not to hold mandatory anti-union meetings or twenty-fifth-hour speeches. The company agreed to allow union representatives access to the plant and set a calendar for when organizers could talk directly to workers. To avoid delays, Smithfield agreed that any union complaints or allegations of company unfair labor practices would be adjudicated on the spot by an NLRB official. Smithfield agreed to accept all rulings by the NLRB, Bruskin said; in return, the union agreed to drop all complaints filed with the NLRB since 1998.

With that, Smithfield dropped the RICO suit. By early morning, word began to spread. At 8 a.m., a group of defendants were having coffee in the café off the lobby, ready to walk over for the trial to start. "There was a rumor in the air that we might have a deal, but nobody had any information," Bruskin said. "Court was supposed to start at 9, so we were eating and waiting. One of the attorneys came in, said that court was canceled and there was a settlement."[27] The reaction? "Holy Shit!"

The attorney seemed pleased but also a little disappointed that after months of preparation they wouldn't get to give their opening statements. "The idea that we were going into open court with no more confidentiality of all the documents. The lawyers were fucking ready to roll," Bruskin recalled. "We wanted to deal, but we wouldn't have minded seeing the company get its butt kicked in court. Still, the goal was to give the people the chance to have the union." Everyone had to go to the courthouse to sign the agreement, and there were a lot of people signing. "I signed the thing, then they took it," Bruskin said, recalling the enthusiasm. "We had the agreement we needed to be able to win. We could have done this a hell of a lot easier and a hell of a lot sooner." Ultimately, he said, it had come down to Joe Hansen. "He was the authority. No decision could be made without the president."

The RICO litigation was dismissed without prejudice, meaning that the lawsuit could be reinstated if the union did not comply with the terms of the settlement. In the settlement, the union signed away ownership of the words "Justice@Smithfield" and agreed to immediately take down all evidence of the campaign from the internet. The website—and all documentation of worker abuse—had to be gone in twenty-four hours. "That was the deal. It had to look like it had never happened," Bruskin said.

Both sides agreed to sponsor a joint program called "Feeding the Hungry," to deliver 20 million servings of protein to the poor over three years.[28] They also agreed not to defame each other, which became a de facto gag order. (As a direct result, both the UFCW and Smithfield Foods declined to comment for this book. The NLRB provided nearly all requested documents under the Freedom of Information Act.)

Bruskin felt that the union came out ahead, winning the concessions from Smithfield it had sought all along. With a level playing field, Bruskin thought they could win. "There was a detailed agreed-on set of rules for the election," he said. "We were confident that with oversight the union would win. We still knew it would be hard. With five thousand workers and the company having access to them every second of the day if they wanted while we had to go and find them. We knew we couldn't take anything for granted."

Bruskin also believes that if Joe Luter III had not retired from his position as CEO, the agreement would not have been possible. "Luter had enormous personal investment in never giving in to the union. . . . As much as he was responsible for the abuse, he was convinced he was the victim of this vicious union and he would never settle. [CEO ] Larry Pope didn't have that personal long-standing animosity. For him, it was just a business decision."[29]

---

After RICO, the campaign returned in full force to the ground game inside the plant. It was a whole new world. Thousands of Latinos had fled or been deported. Organizers felt sucker punched, watching all their hard work—all their an-inch-at-a-time work over the years—drain out across the state, some toward the Mexican border, within a matter of months. It was an unstoppable breach. "Workers just quietly moved on," one organizer said. "You couldn't tell them, stand up for the union. They'd say, 'I'd rather go take my family than be deported.' There's no arguing with that. Some went to smaller companies. A lot just packed up and left and went back."[30]

ICE's approach undermined community support for immigrant workers as well. Busting undocumented workers who were helping save the economically starved towns around the plant was one thing. But federal charges that workers had stolen U.S. citizens' identity? That made people turn away. Union organizers recalled a sense of devastation and loss, of depressed morale and a sense of defeat.

In 2006, 2007, and 2008, between 433,000 and 479,000 Mexican migrants in the United States returned to Mexico each year.[31] In 2008, North Carolina was still home to about 601,000 Hispanics. About half were undocumented, the eighth highest in the country.[32] At the plant, in two years, the racial makeup had gone from being mostly Hispanic to mostly black. In 2006, 46 percent of the five thousand workers were Hispanic, while 41 percent were black, 5 percent were white, and 8 percent were "other." By 2008, the Latino workforce at the plant had dropped from more than half to about 26 percent of the workforce, while blacks had risen to 53 percent and whites to nearly 12 percent; "other" remained at about 8 percent. When Latinos fled to escape federal immigration crackdowns, Smithfield began recruiting African Americans—promising improved working conditions and better pay—to make sure that the first hog got hooked every day and those lines kept moving meat along to the world's tables.[33]

After RICO, the union immediately brought in some hundred organizers, all staying at the Holiday Inn in Lumberton, moving from mostly Spanish-speaking organizers back to African Americans. They now had a list of all workers dating back to 1993; some two thousand workers needed to be contacted. The challenge was enormous, Ludlum recalled. "These were poor people. They were there one week, the next week, 'I ain't got the rent. My aunt kicked me out.' We were constantly chasing people down. And international [the UFCW] was constantly criticizing us. 'Dude, we're doing everything we can.'" What Ludlum wasn't told initially was that his name had specifically come up during the midnight-oil negotiations of the RICO settlement. But he soon found out.

Every day the "action team" met at a small Latino store in Tar Heel to plan the day's action. A portion of the team was still working exclusively on signing up Latinos. Ludlum left the meeting and headed for the plant, where he was spotted by Smithfield management. By the time he got back to the store, a phone call was waiting from the UFCW. "Come back to the office in Lumberton," the supervisor said. When Ludlum got to Lumberton, he recalled, Emily Stewart, then campaigns director for the international union, was waiting to explain that the UFCW had agreed that Keith Ludlum was not to be allowed near the plant in Tar Heel.

Ludlum was stunned. 'It's kind of stupid to allow me to be on the action team and you've cut this deal and nobody's told me anything about it.'" Stewart and her colleague, another supervisor, apologized, Ludlum said. "I said that was fine. I had gotten them to the five-yard line. They could take it into the end zone and take credit as long as we won." They also "bumped Gene [Bruskin] out at the end so he wouldn't get credit."[34] Bruskin declined to comment on the circumstances of his departure. "There was a lot of disagreement about different people's roles," he said. "I was not a UFCW employee. I was still on loan. There were a lot of voices and people were worried [about winning the election]. There were definitely things done that I didn't agree with, but that was all background noise. It was upsetting to me at different points. There were questions about what the union actually did, but the agreement was the right thing to do. I moved on. I got an offer to work at the AFT [American Federation of Teachers]."[35]

Ludlum didn't know whether to be hurt or flattered. "It was so funny that this huge company paid so much attention to one person as part of this big settlement," he said.[36] One insider would later say the union threw Ludlum "under the bus."[37]

With Ludlum and Bruskin gone, new union leadership deemphasized public protests and focused on its ground campaign inside the plant, where they now were allowed to meet directly with workers, as specified in the RICO settlement. The union was allowed equal time to dispute claims by the company that unionizing would be bad for the workers. No one could be fired or intimidated for union support. There would be instant enforcement of the NLRA.

A monthly newsletter, *Gone Hog Wild*, carried stories of safety and wage violations, mistreatment of women, and opinions about the upcoming 2008 U.S. presidential campaign, in which African Americans were keenly interested. The newsletter sought to educate workers: "Federal law gives you the right to form, join or assist a union if you want to. . . . Showing your support for the union is protected under Federal Law."[38] Organizers inside the plant recall sensing a change in the workers, a new receptiveness, like the dawning of a new awareness.

Some credited the Latino walkout in November of 2006, which resulted in demands being met. African Americans had watched as hundreds of workers stood outside chanting, blasting music, and making speeches for two days. They negotiated with the man, got their demands answered, kept their jobs, and went back to work, at least temporarily. "There was a spark that triggered something," one organizer said. "I think the spark came from that walkout. That was the transformative movement for the workforce."[39]

Others believed it was simmering anger among African Americans on account of not getting Martin Luther King Jr. Day off that January of 2007. It was the first concrete thing the black workers had demanded, but the company had refused to listen, even when they walked out. Still others believed it was the power of union leadership in Livestock that set an example for other African Americans in the plant.

Whatever the reason, African Americans and Latinos began to work in solidarity. A Latino worker, following Keith Ludlum's example on the kill floor, where the majority of workers were black, used a felt pen to write "Union Time" on his helmet. Fellow line workers joined him, and the practice spread. Even after Ludlum was fired, union workers met every day before and after work in the back room of a tiny Mexican store, the Tienda La Hacienda, near the plant.

As Martin Luther King Jr.'s birthday rolled around again that January of 2008, union workers passed out flyers saying, "Hold the Date." This time, the company gave workers the day off as a paid holiday. That bolstered confidence.

Unprecedented access meant that union organizers could meet with workers during lunch and in break rooms. Anti-union "captive audience" meetings were limited in number and much more restrained, devoid of the threats that characterized the earlier campaigns.[40]

---

On December 10 and 11, 2008, as the rain fell intermittently, nervous workers stood outside the Tienda La Hacienda in the center of Tar Heel. They held group prayers, videoed each other saying positive things, and put their hands together in the center of a circle, then threw them in the air shouting "Union Time!" The stakes were high. "This plant is the flagship, not only for Smithfield, but for the meatpacking industry. Smithfield is going to determine whether the meatpacking jobs in the future are going to be good union jobs, or going to be low-wage dangerous jobs," Peña said.[41]

The vote count was announced that night: In the carefully monitored vote, workers chose the union by the narrow margin of 2,041 to 1,879, just 162 votes.

Some say that, in the end, national and international politics tipped the scales. One newspaper concluded that the ICE raids sealed the union's victory by shifting the majority population from Latino to black. "The 2007 raids purged the plant of illegal Hispanic workers and left behind a majority of native workers more likely to support unionization," reported the *Charlotte Observer*. "The new black majority proved to be the difference."[42] Others believe that it was Barack Obama's support for the Tar Heel workers that inspired African American workers and led to the victory.

Joe Luter III, sitting in Smithfield headquarters in Smithfield, Virginia, near a portrait of his father, conceded that the fight against the union had been a mistake. "Economically, we should have recognized the union and [avoided] the boycott. . . . A lot of people believe, for some reason, that the company is not unionized, or that we're anti-union, and that's not the case."[43]

On December 11, 2008, workers and union organizers gathered at the Holiday Inn in Lumberton chant "Yes we can!" as word arrives that the union has been voted in. Photo by Ed Panas, *Fayetteville Observer*, December 11, 2008.

While waiting for final confirmation of the vote and victory, workers had gathered at the Holiday Inn in Lumberton. When the phone call came, they screamed, hugging and grabbing each other. Gene Bruskin burst into the ballroom, pumping his fist with a guttural yell. Workers danced, chanting "Si, se puede! Si, se puede!" Bruskin grabbed a microphone. "As soon as I got down here and started meeting with the workers in this plant, I realized we were going to be able to do it," he said. He credited workers by name, one by one, as cheers went up. "I've been doing this for thirty years and I have never met a more courageous group of workers in my entire life."[44]

On July 1, 2009, more than four thousand employees at the Tar Heel plant got a $1.50 an hour raise, the largest pay raise in the history of the plant. Today, UFCW Local 1208 has 3,455 members, assets of $653,831, and sixteen employees and maintains an office near the original trailer in Tar Heel.[45] "It was like it was meant to be and nothing was going to stop it, like God has his hand in everything," Ludlum said. "The path was crooked and winding, but at the end the finish line was the victory for the workers."[46]

I n 2011, when I first met Sherri Buffkin, she wasn't looking much like the hero who'd been presented to Congress. She'd had two heart attacks and a long bout of pneumonia, and had broken her back when a truck ran her off the road. Smithfield blackballed her and drove her into bankruptcy in February of 2000, she said.[1] The only work she could find was selling cars more than an hour away. During our interview, she sat on her worn couch, with *Sex and the City* playing soundlessly on the TV nearby. She wore sweat-pants and a T-shirt, her hair pulled back in a ponytail, and played with her large dog Samo with a torn football.

Buffkin had burned what remained of her legal papers about five years before; she still remembers that night. It was around dusk. Sparks shot up as she threw papers into a burn barrel; smoke drifted up into the North Carolina sky. Her husband asked if there was anything he could do. She said, "All I need is a drink." He brought her a pitcher of margaritas and said, "If that pitcher don't do it, I'll bring you another." Buffkin told him, "I promise you, baby. We'll be done." Buffkin watched as curling chars of gray twirled away, carrying pain and hurt. "It's over. It's done. They can't hurt us anymore." One last box of bad memories.

Buffkin shook off the memory, let the dog out, and picked up a cigarette from a pack on the wooden bench on the front porch, then coughed and coughed in the cold air before going back inside. "Smithfield owns this state. They own this county," she said in her deep North Carolina drawl. "Smithfield spent millions in attorneys' fees pursuing me and trying to shut me up once I started talking."[2]

---

That same year, workers convinced Keith Ludlum to come back to the union. They were unhappy with the local managers the UFCW had put in charge after the union won in 2008, Ludlum said. The union office had been moved to Lumberton, a half hour away, making it difficult for people to go.

That May, when union workers elected Ludlum as president of the local, the first thing he did was rent the old Bridge's Barbecue building right in Tar Heel and move the union office back to its hometown.

"There was no respect inside the plant for the union reps. Management was playing games. So I told the reps: Respect the worker on the line. The minute you got a problem, you stop," Ludlum said. Supervisors were still cursing out workers and telling them that in a right-to-work state they didn't have to join the union. Ludlum was especially outraged about bathroom breaks. He told reps to enforce the rules that workers could use the bathroom any time they wanted and could stay as long as they needed. "Right to work does hurt you, because you've got all the freeloaders along for the ride, enjoying the benefits, but they don't have to sign on," Ludlum said.[3] In right-to-work states, even if a union is voted in, workers have the right to refuse to join the union or pay dues, yet they benefit from union negotiations.

Ludlum said that he got at least forty people reinstated after they were fired, most with back pay, and had numerous suspensions removed. As a result, he said, supervisors became wary about disciplining union members without cause. "We sat down with HR managers once a week and went over cases we disagreed with. Most of the time, they didn't know what was really going on." In one case, Ludlum said, he exposed sexual harassment. In another he had a supervisor removed until evidence could be gathered against a worker. "I held the company to a very high standard."[4]

Every Wednesday, Ludlum, local union officials, and union stewards stood in the hallways handing out notices and flyers during shift changes. In December of 2014, Ludlum was reelected as president of Local 1208. He had successfully negotiated two contracts with increased benefits to workers, he said. In his first two years, he said, he increased membership in the union at the plant from 50 percent to 80 percent, though these figures could not be confirmed because the union declined to comment for the book, citing the RICO gag order.

Ludlum began to set his sights on unionizing other meatpacking plants, focusing on Mountaire Farms, a chicken slaughterhouse with about two thousand workers, as well as speaking out frequently in newspaper articles, on radio shows, and on panels. In March, the union filed charges with the NLRB alleging that Mountaire management had threatened to arrest and fire workers and done illegal surveillance. By June, seven hundred workers had signed union cards. After the company hired Labor Relations Institute, a professional anti-union consultant, the pace of sign-up slowed considerably.

Mountaire was employing the same tactics used by Smithfield, Ludlum told an audience at Duke University on February 17, 2014, during an "Organize the South" rally. "So what's different? We know their play book. We've got a different dance and we learned it at Smithfield." Unionization, he told them, is the only way to erase the disparity between the 1 percent and the rest of the country.[5] "We've declared that organizing the South is the only option we have. We have 5,000 in Tar Heel, another 2,000 at Mountaire, another 2,500 in Perdue and that's just southeast North Carolina. We have workers in Raleigh and Durham and Charlotte and everybody will be coming together."[6]

Ludlum even got Danny Glover to come and speak to Mountaire workers. Glover took the red-eye from San Francisco. On August 6, 2014, wearing a yellow UFCW T-shirt with "Local 1208" on it, he stood in the humid heat outside the entrance to the Mountaire plant in Robeson County. "We are the union, mighty, mighty union," the 68-year-old actor chanted. Referring to Glover's four hit films with Mel Gibson, Ludlum called out to carloads of workers: "Hey, we brought you a lethal weapon! Don't let them intimidate you!"[7]

Details are not available, but in the background trouble was brewing for Ludlum regarding his handling of union funds. Ludlum says he ruffled feathers with his hard-driving personality and had fired union staff who disagreed with his tactics. In September of 2014, the Department of Labor seized the local's computerized accounting records for an audit. A year later, Ludlum said, everything was returned and no charges were brought. Still, by March of 2015, the UFCW had placed the local in trusteeship and replaced Ludlum with UFCW staffers from the International, not the local. "They said I was misappropriating funds. But they knew we were doing stuff wrong. They allowed it to go on. It wasn't a secret," he said. According to Ludlum, no one trained the staff in how to handle funds, and when there were immediate union needs that had to be paid for, he would often not get approval. There is some evidence that staff complained to Ludlum's superiors.[8] As of June 2016, Ludlum hadn't been able to find work. The trusteeship has been removed, and new leadership is in place.[9]

By 2016, Sherri Buffkin had climbed out of the hole Smithfield had put her in, she said. She was still married to Davie, who still works at the plant in sanitation. She'd had heart surgery. She'd paid off her house with the Smithfield sexual harassment lawsuit settlement and paid her daughter's way through nursing school. She'd worked her way up to top salesperson at John Donoghue Automotive in Whiteville, less than twenty minutes from home. She'd gotten down from size 12 to size 4 and had cut way back on her drinking. "Not too bad for almost 47," she texted, with a picture of herself in a sleeveless pink dress. "From hell and back. . . . They might have gotten me down for a while but I'm back on top. . . . Just a lot smarter/wiser."

Buffkin never imagined how great the damage of going up against Smithfield would be, its effect on family, friends, and neighbors in a county where everyone knows everyone. "People in Bladen County don't understand," she said in 2011. "They never read the court's decisions. They don't know that Judge West said I was telling the truth about every single thing."

Buffkin said she lost more than $10,000 from the settlement money that Smithfield had held back to ensure confidentiality. The reason? She punched Susie Jackson in the nose in the foyer of their church a few years after Jackson testified against her. "You know me and my temper." The charges were dismissed, but "it cost me. They took the money from me for beating her up. She ended up in the hospital. I defended myself real good. . . . I had more than seventy employees and none of them would go to court and testify against me. I gave her a job and made her a supervisor and she was the only one who would testify. . . . She thought she could get ahead and now she's gone."

When Susie Jackson got cancer, Buffkin recalled, she phoned Sherri to make her peace. "She said, 'I'm sorry for everything that happened and for everything I said. Will you come and see me?' I said, 'I forgive you, but we will never be friends. Rest in peace, but I have no desire to see you.' . . . Her sister called to tell me she died."

Buffkin was upset to hear that Ada Perry had died in 2015. "Granny's gone? I hate that. Cause that crazy-ass woman? She called to thank me. They were giving her back pay and they were offering her job back. All kinds of crazy mess. Granny's Granny. You got to love her. I hate to hear she passed."

Even after years of reflection, Buffkin still can't really answer the big question: why did she get fired from Smithfield in the first place? "I think they knew I was going to tell the truth. Everything had been building up. I'd been telling them, 'You can't do this.' When Larry fired me, he told me, 'I don't trust you anymore,' cause I told him, 'You need to quit this shit.'"

She never intended to be the catalyst for the unionization of the world's largest pork slaughtering house. But without her? No way. "How close do you think the union would have come to getting in?" she said. "There was no one else in a position of power who was willing to speak the truth. . . . Smithfield wasted millions of dollars fighting me in court. And because this one little girl, this one country-ass woman, stood up and told the truth. That's my story. And guess what. The people who didn't listen weren't my friends to begin with.

"Joe Luter? Several times I met him when he didn't know who I was, but he's known who the hell I was for a long time now. This little country white ass, he's known for a long time. If he'd just come to me then, he'd have saved himself a whole lot of trouble.

"I'm one of the few people on the face of this earth who fought them and won."[10]

---

By June of 2016, Larry Johnson looked like he was enjoying being rich. I found him outside his large brick home in Elizabethtown, where he was washing down one of four vehicles in front of a stand-alone three-car garage (in addition to the two-car garage that was part of the house). I asked him why he left Smithfield Packing. "I'll be real honest. I'm kind of a people person. New management took over and I didn't fit into the mold," he said. "They paid me well to go away, so it worked out, but I really can't talk."

Johnson suggested that Jere Null might be more willing to talk. "I'm not going to say he's more bitter than I am, and I'm not really bitter, but I did make a promise not to disparage the company and I have a pension. I wouldn't want to jeopardize that. . . . The Luter family, from the time I met Joe Luter III, it was a wonderful experience, until 2009. To make a long story short, under Joe III, each company operated independently, but as time went on, we were competing within ourselves and there was a whole lot of consolidation. The whole structure of the company has changed. Joe IV is out of there. He's got his own business with tacos and condos and God knows what. Jere Null would be the one to get hold of. He's more articulate and more knowledgeable of the inner workings."

Johnson recalled when he was a union rep for the UFCW in Wisconsin in a closed shop when workers went on strike. "I seen the handwriting on the wall. The days of me making—as a layperson—$14, $15 an hour in the seventies were over. Looking forward, I went into management. I had nothing against the union at all. When it came to Tar Heel? An organizing group and an established union are different entities. Organizing is hostile. They'll do anything. The president of the union over there, he was one of the major organizers, but being the person that he was. . . . Anyway, I hear he's gone and I hear they got an interim over there and things are good.

"I'd rather manage a plant that's unionized. Of course, it's easier in a right-to-work state because there's a portion of the plant that's union and a portion that isn't. I always argue, you all got to be all in. Otherwise, it's not going to be strong.

"There's a transition going on in the company, John Morrell Group is reorganizing. I'll tell you why I can't talk to you. There were a lot of things said. A lot of accusations on both sides. Judge West made everyone out to be liars, said everybody in management was lying. I don't know if I want to go through that again.

"When I left Smithfield Packing, the rumor mill was because I embezzled all this money and I had a yacht in the Bahamas. I'm working for another pork company now. I'm 68 and have no intentions of retiring. My philosophy is I serve the workers. If you spend any time around here and bring my name up, I can almost guarantee you they'll say they wished I was back. . . . I'll call Jere and tell him to call you. He still has a residence in Smithfield, still married to Anne. He's working for Prestage now, looking for a site for a big plant." Johnson went back to hosing down his cars outside his house in the hot sun with no trees and no shade.[11]

Jere Null didn't return messages left on his voicemail at Prestage Farms, Inc., in Clinton, North Carolina. Null was not listed on the website under corporate positions, but he's quoted in newspaper articles about the company's search for a site for a new plant in Iowa. The article identifies him as the chief operating officer, normally second in command.

Prestage Farms is one of the top five pork producers in the nation. In March of 2016, they announced they would open their first pork-processing plant in Mason City, Iowa. With Prestage as the largest property taxpayer in town, the city would see $3 million in revenue, the largest financial impact in their history. Iowa offered roughly $13.5 million in incentives to attract the $240-million 650,000-square-foot plant.[12] "I'd be a complete idiot to tell you there won't be an odor associated with any congregation of pork in the world, but I don't think it will be something perceived to be a problem in Mason City," said Ron Prestage, DVM, third son of the company's founder, Bill Prestage.

By April, protesters were gathering at city hall, concerned about negative environmental and economic impacts. A Facebook page was launched: "The People vs. Prestage in Mason City."[13] Posts included a Waterkeeper Alliance report called *Fields of Filth: Landmark Report Maps Feces-Laden Hog and Chicken Operations in North Carolina*. One asked, "How exactly will this be a competitive market, when the big businesses buy out all the contract farms, push out all the smaller family farms and put other markets out of business?"[14]

When the city rejected the incentives package in May, Prestage proposed moving to Sioux City, promising a $264 million plant. The company had been contacted by nearly twenty Iowa communities interested in the project and its estimated two thousand jobs. Ron Prestage was obviously angry when he accused Mason City residents of being racist and called protesters "kooks."[15]

On July 5, Prestage announced that Wright County, Iowa—not Sioux City—would be the location for its new state-of-the-art pork-processing facility. By mid-July, the Wright County Board of Supervisors was rezoning a parcel of land for the plant. Protesters held signs that said, "Wake up and smell the CAFOS." The plant is expected to begin with one shift, then ramp up to two, eventually processing up to about 24,000 hogs per day. Initially, there will be a hog shortage, but as new factory farms open, the supply of young hogs will fill the vacuum.[16]

---

By 2014, Joe Luter was spending time in Palm Beach, with his third wife, Karin, and his daughter, Erika, a teenager. The annual "A Garden in the Wild" dinner dance at the Palm Beach Zoo & Conservation Society on January 31 honored Karin as the inaugural winner of the Gala Stewardship Award. Luter had been chair the previous three years. Karin was called "a shining star in the history of Zoo galas" and a "quintessential leader of this event." A plaque was hung in her honor as "an inspiration for others to act on behalf of wildlife and the natural world for generations to come." Luter and his wife were both on the Chairman's Committee and listed as generous donors.

The two live on the water in a 7,547-square-foot four-bedroom seven-bath home in Boca Raton, appraised at $9,582,517 in 2015. The two-story home with black shutters is framed by large palm trees and surrounded by high hedges. A photo gallery at pattrickmcmullan.com showed sixty-two photos of Karin at a Christmas ball. She was a regular at the annual Christmas Cheer holiday party hosted by heiress Anne Randolph Hearst and her husband, novelist Jay McInerney. Their "Dinner in the Country" event at their second home in the Pine Creek Sporting Club in Okeechobee — with tables set under mossy trees with hanging candle lanterns — was featured in *Palm Beach Entertaining*.[17]

---

Gene Bruskin never met Joe Luter; in fact, he never even laid eyes on him. But he feels like he knows what made him tick. "I feel like he was an angry old man who had made his fortune and resented that these lowly workers were going to disrupt his vision of a non-union plant. He built it from scratch. It was his crown jewel. He was determined these sons of bitches were not going to have their way." Bruskin is now retired, but he hasn't lost his passionate commitment to the rights of workers or his rebellious spirit. While he built on a foundation laid by those who came before, he was indubitably the brilliant mind behind the ultimately successful union campaign.

Bruskin is a charming man, somewhat rumpled, with a deeply lined asymmetrical face and thick glasses. At first, he doesn't seem like someone able to take on and defeat pork-industry magnate Joe Luter. But he is a pit bull when it comes to social justice, unafraid to use his teeth when workers have been illegally stripped of their constitutional rights. A veteran union man, his coup d'état was certainly salvaging — as well as reinventing — the flailing union fight in Tar Heel.

It was the challenge of a lifetime, even for Bruskin. He certainly had extensive help from his teams in Washington and North Carolina and his UFCW bosses, but the antiwar activist — and former United Steelworker — was the field general. His Justice@Smithfield campaign went national as he bypassed the toothless NLRB board. Within three years of Bruskin's coming aboard, the sixteen-year bitter and violent battle between the UFCW and Smithfield Foods over the company's largest plant finally came to an end.

During a 2016 visit at his bungalow in the heart of Silver Spring, Maryland, Bruskin ranged from gleeful to deadly serious, from disarming to somewhat alarming in his radical notions of what is required to beat back corporate greed. Gag orders be damned — he doesn't care. "The story needs to be told, and since the union hasn't been willing to do it, I am," Bruskin said. "I feel like it's important for people to know. When we have a victory, we need to talk about it, study it, and it needs to be part of us moving forward."

Remnants of the union campaign against Smithfield are tucked into boxes in the basement of Bruskin's home, probably illegally, he joked as he led the way down a narrow staircase into a concrete-floored room with open shelves on one side and a washer and dryer on the other. He held up the bright yellow — once ubiquitous — T-shirt bearing the slogan that became synonymous with Smithfield abuse: Justice@Smithfield. "If I wear this and walk down the street, Smithfield could land in a helicopter and grab me for wearing an illegal slogan and put me in prison," he laughed.

At the height of the campaign, an internet search of "Justice@Smithfield" would have yielded dozens of hits, all with damning information about Smithfield Foods, including hundreds of documents, photographs, recordings, videos, and evidence of wrongdoing. In 2016, a search led to half a dozen innocuous sites unrelated to the campaign. Even one that led to smithfieldjustice.org, linked to ufcw.org, contained no information about the campaign. It's as if Justice@Smithfield never existed. News releases were altered or removed from both union and company websites. Invaluable historical evidence was destroyed, and Bruskin wants to fill in the lost record, no matter the cost. "What are they going to do? Arrest me? I would galvanize thousands to protest and demand my release," he joked. "A Free Gene campaign would result."

In his classic black-and-white-tiled bathroom, Bruskin has a photo of himself shaking the hand of President Barack Obama. In Bruskin's office is a framed watercolor poster depicting the varying shades of skin color united under his campaign. Under the faces, handwritten as part of the poster design: "In 2008, after 14 years of continued struggle, organizing and campaigning, 5,500 low-income workers won union representation under the United Food and Commercial Workers at the world's largest pork processing plant in Tar Heel, North Carolina. . . . Local 1208. . . . It's all about the workers," Bruskin said.[18]

---

Keith Ludlum lives a world away in rural Bladenboro, in the trailer he's had for years, next door to his dad. There, on a June afternoon in 2016, Keith's father greeted me with a warm handshake and a wary but friendly look, standing on the clean, concrete pad of a driveway, with large, rectangular chicken barns looming behind him farther down the sandy road near the intersection of Highway 131 and Highway 41. "Keith's home," he said, gesturing toward cars parked outside the double-wide some distance away in the adjacent lot.

At the end of another long sandy driveway, in Ludlum's large backyard, a dozen small white chickens huddled in the shade. The two-by-four handrails to the deck were made of unfinished wood, roughly nailed together. The deck was a disorganized collection of tools, tables, and chairs all baking in the hot North Carolina sun. When I knocked on the door, a Latina woman opened it, was gone for some time, then returned to say that Ludlum was asleep. When he got up, we agreed to talk the next day. When I returned, he told me about his family. His ex-wife lives nearby, and their boys are grown. He has a six-month-old child with Dilcia, the woman who answered the door. Dilcia's two older girls were shy and sweet; Dilcia took them fishing in the pond so we could talk.

Keith cashed out his 401k and keeps the children at night so Dilcia can work at the plant. Ludlum believes it's only a matter of time before the union workers vote him back in as president and ask him to come back. He said the UFCW sent a letter to all the union members saying that he had "misappropriated funds." But every year, he said, his superiors in the union had given him an award for being the fastest-growing local. "When somebody is telling them you stole money and did things wrong, how do I make people believe it's not true?" "I said to the union, 'Why now? I've been doing it the same since day one. I did that all the time, go to the board after I spent the money. Why kick me out now? Just correct it and move on.' I'm one hell of an organizer. The company has been getting away with murder since they took me out."

When Ludlum started to remember, reliving his victories, he livened up, flipping his head back and laughing with an open mouth. His blue eyes flashed, full of pleasure in telling the old stories. "It's winning the little battles," he said. "That's what keeps you going through the whole war."[19]

# NOTES

## Introduction

1. Minxin Pei, "The Real Reason behind Shuanghui's Purchase of Smithfield," *Fortune*, June 4, 2013, http://fortune.com/2013/06/04/the-real-reason-behind -shuanghuis-purchase-of-smithfield/.

2. "Food Fight: Meatpacking in the U.S.: Still a 'Jungle' Out There?," *NOW*, PBS, week of Dec. 15, 2006, accessed Aug. 5, 2017, http://www.pbs.org/now/shows/250 /meat-packing.html.

3. "The Price of Pork," *Chicago Tribune*, Aug. 3–8, 2016, http://www.chicagotribune .com/news/watchdog/pork/.

4. U.S. Congress, *Congressional Record: Proceedings and Debates of the U.S. Congress*, 109th Congress, Feb. 17, 2005, Congressional Record—Senate, vol. 151, pt. 2, Assembly Joint Resolution No. 87. p. 2749, https://books.google.com/books?id=M7 EWi89eCA4C&pg=PA2749&dq=how+many+workers+suffer+illegal+retaliation +for+supporting+unions&hl=en&sa=X&ved=0ahUKEwiws223azSAhWh0YMKHZ wRBSgQ6AEIIjAB#v=onepage&q=how%20many%20workers%20suffer%20 illegal%20retaliation%20for%20supporting%20unions&f=false.

5. Noam Scheiber, "Rule to Require Employers to Disclose Use of Anti-Union Consultants," *New York Times*, March 24, 2016, https://www.nytimes.com/2016/03/24 /business/economy/union-labor-regulation-consultant-relationships.html?_r=0.

6. Wilma Liebman, former chair, NLRB, telephone interview by author, Feb. 11, 2017.

7. Wilma Liebman, "U.S. Trade Unions and German Subsidiaries," *Soziales Recht* (Dec. 2016): 143–157.

8. Paul Krugman, review of *Capital in the Twenty-First Century* by Thomas Piketty, *New York Review of Books*, May 8, 2014, http://www.nybooks.com/articles/2014/05 /08/thomas-piketty-new-gilded-age/.

9. Ibid.

10. Nicholas Kristof, "An Idiot's Guide to Inequality," *New York Times*, July 23, 2014, https://www.nytimes.com/2014/07/24/opinion/nicholas-kristof-idiots-guide -to-inequality-piketty-capital.html.

11. Liebman, "U.S. Trade Unions and German Subsidiaries," 143–157.

12. Paul Krugman, "State of the Unions," *New York Times*, Dec. 24, 2007, http:// www.nytimes.com/2007/12/24/opinion/24krugman.html.

13. "The American Middle Class Is Losing Ground: No Longer the Majority and Falling Behind Financially," Pew Research Center, Dec. 9, 2015, accessed Aug. 5, 2017, www.pewsocialtrends.org/2015/12/09/the-american-middle-class-is-losing-ground/.

14. Liebman, "U.S. Trade Unions and German Subsidiaries," 143–157.

15. John N. Raudabaugh, "The Raudabaugh Report: NLRB Productivity Cost Analysis FY1980–FY2014," http://www.nrtw.org/the-raudabaugh-report-nlrb -productivity-cost-analysis.

16. Katherine Peralta, "North Carolina's Union Membership Rate Is the Lowest in the Country," *Charlotte Observer*, Jan. 28, 2015, http://www.charlotteobserver.com /news/business/article9285041.html.

17. Richard B. Freeman, "What Can We Learn from the NLRA to Create Labor Law for the Twenty-First Century?," *A.B.A. J. Lab. & Emp. L.* 26 (2011); 327, 330.

18. Paul Weiler, "Promises to Keep: Securing Workers' Rights to Self-Organization under the NLRA," *Harvard Law Review* (June 1983): 1769.

19. Liebman, interview, Feb. 11, 2017.

20. Danielle Paquette, "Donald Trump and Labor Unions Don't Always Get Along—but They Did Today," *Washington Post*, Jan. 32, 2017, https://www .washingtonpost.com/news/wonk/wp/2017/01/23/donald-trump-tells-union-leaders -he-wants-to-hire-americans-to-rebuild-america/?utm_term=.01376905ec98.

21. Steven Greenhouse, "What Unions Got Wrong about Trump," *New York Times*, Nov. 26, 2016, https://www.nytimes.com/2016/11/26/opinion/sunday/what-unions -got-wrong-about-trump.html.

22. Liebman, "U.S. Trade Unions and German Subsidiaries," 143–157.

23. Liebman, telephone interview by author, Feb. 4, 2017.

24. Greenhouse, "What Unions Got Wrong about Trump."

25. Liebman, interview, Feb. 11, 2017.

26. Greenhouse, "What Unions Got Wrong about Trump."

27. Ibid.

28. Liebman, interview, Feb. 4, 2017.

29. Liebman, interview, Feb. 11, 2017.

30. Nelson Lichtenstein, *State of the Union: A Century of American Labor* (Princeton, NJ: Princeton University Press, 2002), 16.

31. Gene Bruskin, telephone interview by author, March 9, 2016.

## Chapter 1

1. U.S. Department of Commerce, U.S. Census Bureau, accessed Oct. 19, 2017, https://factfinder.census.gov; North Carolina Department of Commerce, ACCESSNC, North Carolina Economic Data and Site Information, County Profile, Bladen and Robeson Counties (NC), accessed Oct. 19, 2017, http://accessnc.nccommerce.com /DemoGraphicsReports/; Lynn Waltz, "Slaughterhouse '05," *Portfolio Weekly*, July 5, 2005. (Note: In 2005 in Bladen County, one in five lived in poverty.)

2. U.S. House of Representatives, "Hidden Tragedy: Underreporting of Workplace Injuries and Illnesses," June 2008, Majority Staff Report, U.S. House Committee on Education and Labor, accessed Aug. 5, 2017, https://www.bls.gov/iif/laborcomm

report061908.pdf; *Meatpacking*, United States Department of Labor, Occupational Safety and Health Administration, Safety and Health Topics, accessed Aug. 5, 2017, https://www.osha.gov/SLTC/meatpacking/; "Meat Packing Plants Have the Highest Rate of Repeated-Trauma Disorders," *Economics Daily*, Bureau of Labor Statistics, U.S. Department of Labor, Aug. 5, 1999, accessed Apr. 2, 2017, https://www.bls.gov /opub/ted/1999/aug/wk1/art04.htm.

3. Frank Maley, "The Meat of the Matter," *BusinessNC*, Apr. 2008, http://business nc.com/category/2008-04/.

4. Kimberly Pierceall, "Hog Sales Drag Down Smithfield Foods' 2015 Profit," *Virginian-Pilot*, March 29, 2016, http://pilotonline.com/business/stocks/hog-sales -drag-down-smithfield-foods-profit/article_09e09786-47d3-5f53-86d4-bc2b42e6d5f6 .html; Tara Bozick, "Smithfield Foods Reports Record Operating Profit in First Half of 2016," *Daily Press*, Aug. 17, 2016.

5. Smithfield Foods, "Our Operations," accessed Aug. 5, 2017, http://www.smith fieldfoods.com/about-smithfield/our-operations.

6. Rogers Dey Wichard, *History of Lower Tidewater Virginia*, vol. 3 (New York: Lewis Historical Publishing Company, 1959), https://archive.org/details/historyof lowertio3whic; Barbara Young, "Smithfield Foods' 21st Century Approach to Food Business," *National Provisioner*, June 2001, cover story.

7. Col. E. M. Morrison, "A Brief History of Isle of Wight County, Virginia," *History of Isle of Wight County — 1608–1907* (Isle of Wight County [Virginia] Historical Society, n.d.), accessed Aug. 5, 2017, http://www.iwchs.com/IWCHistory.html.

8. "Smithfield, Virginia: A Brief Historical Overview," Smithfield & Isle of Wight Convention & Visitors Bureau, accessed Aug. 5, 2017, http://www.genuinesmith fieldva.com/a-brief-history-of-smithfield.html.

9. Amy Brecount White, "Ham-burg, Virginia: In Fragrant Smithfield, a Taste of Life in the Pig City," *Washington Post*, May 29, 2002.

10. Richard Ernsberger Jr., "The Ham Man," *Virginia Living*, Sept. 25, 2009, http://www.virginialiving.com/culture/the-ham-man/.

11. Karl Rhodes, "The Bacon for Smithfield Foods," *Richmond Times-Dispatch*, Sept. 1, 1987.

12. P. D. Gwaltney, Jr., and Company, Inc., and United Packinghouse Workers of America, CIO. NLRB Case No. 5-R-2256, Supplemental Decision and Order, June 25, 1947.

13. Ibid.

14. Ibid.

15. Smithfield Packing Company, Incorporated and United Packinghouse Workers of America, CIO, NLRB Case No. 5-RC-1597, May 24, 1955.

16. Transcript: Before the National Labor Relations Board, Region Eleven. In the Matter of: The Smithfield Packing Company, Inc. — Tar Heel Division and United Food and Commercial Workers Union Local 204, AFL-CIO, CLC. Case numbers 11-CA-15522, 15634, 15666, 15750, 15871, 15986, 16010, 16161, 16423, 16680, 17636,

17707, 17763, 17824, and 11-RC-6221. Before Honorable John West, Administrative Law Judge. Provided by the NLRB in response to FOIA in searchable PDF format. (Hereafter "NLRB Transcript.")

17. Carol Lichti, "The Man Behind a $50 Million Naming-Rights Offer and a $6 Billion Company," *Inside Business: The Hampton Roads Business Journal*, Feb. 4, 2002, http://pilotonline.com/inside-business/news/other/the-man-behind-a-million-naming -rights-offer-and-a/article_99fda7f9-f787-51c9-8cb3-dbc8be79f3dd.html.

18. Rhodes, "The Bacon for Smithfield Foods."

19. Lon Wagner, "Smithfield Foods CEO Joe Luter: No Apologies," *Virginian-Pilot*, Nov. 19, 1995.

20. Ernsberger, "The Ham Man."

21. Rhodes, "The Bacon for Smithfield Foods."

22. Wagner, "Smithfield Foods CEO Joe Luter: No Apologies."

23. Rhodes, "The Bacon for Smithfield Foods."

24. Wagner, "Smithfield Foods CEO Joe Luter, No Apologies."

25. Ernsberger, "The Ham Man."

26. Rhodes, "The Bacon for Smithfield Foods"; Dale Miller, "Straight Talk from Smithfield's Joe Luter," *National Hog Farmer*, May 1, 2000, http://www.nationalhog farmer.com/mag/farming_straight_talk_smithfields.

27. Stephen J. Lynton, "Pork Baron Brings Home Bacon by Trimming Smithfield's Fat," *Washington Post*, Feb. 7, 1983.

28. Rhodes, "The Bacon for Smithfield Foods."

29. Ibid.

30. "Carroll's Foods, Inc.—Company Profile, Information, Business Description, History, Background Information on Carroll's Foods, Inc.," referenceforbusiness.com, accessed Aug. 5, 2017, http://www.referenceforbusiness.com/history2/26/Carroll-s -Foods-Inc.html#ixzz4GNuiFrme.

31. Waltz, "Slaughterhouse '05"; Robert F. Kennedy Jr., draft of foreword to *Righteous Porkchop: Finding a Life and a Good Food beyond Factory Farms* by Nicolette Hahn Niman, provided by Robert F. Kennedy Jr. to author in 2004.

32. Wagner, "Smithfield Foods CEO Joe Luter: No Apologies"; Bill Geroux, "Smithfield Backs Its 'Foods' but Some Balk at Meatpacker's Nomination for Award," *Richmond Times-Dispatch*, Feb. 1, 1998.

33. Lon Wagner, "Smithfield Foods CEO Joe Luter: No Apologies."

34. White, "Ham-burg, Virginia."

35. Rhodes, "The Bacon for Smithfield Foods."

36. Ernsberger, "The Ham Man."

37. Roger Horowitz, *"Negro and White, Unite and Fight!"* (Champaign: University of Illinois Press, 1997), 253.

38. "UFCW Meat Packing & Food Processing: History; A History of Organizing for Power," UFCW.org, accessed Aug. 4, 2017, http://www.ufcw.org/meat-packing/.

39. Halpern, *Down on the Killing Floor*, 39–44; Rick Halpern and Roger Horowitz,

*Meatpackers: An Oral History of Black Packinghouse Workers and Their Struggle for Racial and Economic Equality* (Monthly Review Press, 1999), 11–12, 27.

40. Horowitz, *"Negro and White, Unite and Fight!,"* 245.

41. "Meatpacking: Color It Green," *Newsweek*, March 8, 1965, 76.

42. Horowitz, *"Negro and White, Unite and Fight!,"* 247.

43. Leon Fink and E. Alvis, *The Maya of Morganton: Work and Community in the Nuevo New South* (Chapel Hill: University of North Carolina Press, 2003), e-book, 10.

44. Wilson J. Warren, *Tied to the Great Packing Machine: The Midwest and Meat-packing* (Iowa City: University of Iowa Press, 2007), 24–26.

45. Horowitz, *"Negro and White, Unite and Fight!,"* 247.

46. Warren, *Tied to the Great Packing Machine*, 42–43.

47. "Smithfield Foods," *Richmond Times-Dispatch* (city edition), Aug. 21, 1985, NewsBank record 8501010317; "1 Person Arrested in Cudahy Strike," *UPI News Track*, Jan. 5, 2987, NewsBank record 1987-091-05-8396536821200.

48. Rhodes, "The Bacon for Smithfield Foods."

49. Ibid.

Chapter 2

1. Peralta, "North Carolina's Union Membership Rate Is the Lowest in the Country."

2. Sherri Buffkin, telephone interview by author, Aug. 13, 2011.

3. William O'Neal, "The Emergence of the Crop-Lien System in Eastern North Carolina," paper prepared for Dr. Donald H. Parkerson, HIST 4000, East Carolina University, Greenville, NC, Spring 2010, http://thescholarship.ecu.edu/handle /10342/3795.

4. David Walbert, "Town and Villages," *Learn NC: K-12*, http://www.learnnc.org /lp/editions/nchist-antebellum/5332.

5. Frederick Law Olmsted, *A Journey in the Seaboard Slave States; With Remarks on Their Economy* (New York: Dix and Edwards, 1856), 338–351.

6. Outland, Robert B., III, *Tapping the Pines: The Naval Stores Industry in the American South* (Baton Rouge: Louisiana State University Press, 2004), 5–6.

7. David Walbert, "Distribution of Land and Slaves," *Learn NC*, http://www .learnnc.org/lp/editions/nchist-antebellum/5347.

8. "Lynchings: By State and Race, 1882–1958," University of Missouri Kansas City, Faculty Project; statistics provided by the Archives at Tuskegee Institute, http://law2 .umkc.edu/faculty/projects/ftrials/shipp/lynchingsstate.html.

9. "Labor and Unions in North Carolina Textile Mills," *Greensboro Truth & Reconciliation Commission Final Report*, 66–98, http://www.greensborotrc.org/pre1979 _labor.pdf.

10. Ibid.

11. Ibid.

12. Ibid.

13. Ibid.

14. Ibid.

15. Ibid.

16. Ibid.

17. Brent D. Glass and Wiley J. Williams, "North Carolina Organized Labor: The Modern Era," in *Encyclopedia of North Carolina*, edited by William S. Powell (Chapel Hill: University of North Carolina Press, 2006), accessed Aug. 5, 2017, http://www.ncpedia.org/labor-unions-part-5-north-carolina.

## Chapter 3

1. Rhodes, "The Bacon for Smithfield Foods."

2. NLRB Transcript.

3. "Plant Wants OK to Kill 1 Million More Hogs," *StarNewsOnline*, July 26, 2002, http://www.starnewsonline.com/news/20020726/plant-wants-ok-to-kill-1-million-more-hogs.

4. David Kirby, *Animal Factory: The Looming Threat of Industrial Pig, Dairy and Poultry Farms to Humans and the Environment* (New York: St. Martin's, 2010), 88.

5. Melanie Sill, Pat Stith, and Joby Warrick, "Boss Hog: The Power of Pork; North Carolina's Pork Revolution," *News & Observer*, Feb. 19–28, 1995, http://www.pulitzer.org/winners/news-observer-raleigh-nc-work-melanie-sill-pat-stith-and-joby-warrick.

6. Pat Stith and Joby Warrick, "Murphy's Law: For Murphy, Good Government Means Good Business," *News & Observer*, Feb. 22, 1995, http://www.pulitzer.org/winners/news-observer-raleigh-nc-work-melanie-sill-pat-stith-and-joby-warrick.

7. Miller, "Straight Talk from Smithfield's Joe Luter."

8. Christopher Leonard, *The Meat Racket: The Secret Takeover of America's Food Business* (New York: Simon & Schuster, 2014), 167–168.

9. David Barboza "Goliath of the Hog World: Fast Rise of Smithfield Foods Makes Regulators Wary," *New York Times*, Apr. 7, 2004, http://www.nytimes.com/2000/04/07/business/goliath-of-the-hog-world-fast-rise-of-smithfield-foods-makes-regulators-wary.html?pagewanted=1&src=pm.

10. Leonard, *The Meat Racket*, 166.

11. Ibid., 5.

12. "Key Industries: Hog Farming," *Learn NC*, http://www.learnnc.org/lp/editions/nchist-recent/6257.

13. Joby Warrick and Pat Stith, "Corporate Takeover," *News & Observer*, Feb. 21, 1995, http://www.pulitzer.org/winners/news-observer-raleigh-nc-work-melanie-sill-pat-stith-and-joby-warrick.

14. David Barboza, "Goliath of the Hog World: Fast Rise of Smithfield Foods Makes Regulators Wary," *New York Times*, Apr. 7, 2004, http://www.nytimes.

com/2000/04/07/business/goliath-of-the-hog-world-fast-rise-of-smithfield-foods
-makes-regulators-wary.html?pagewanted=1&src=pm.

15. Miller, "Straight Talk from Smithfield's Joe Luter."

16. Joby Warrick and Pat Stith, "Corporate Takeover," *News & Observer*, Feb. 21, 1995.

17. Robert F. Kennedy Jr., telephone interview by author, Oct. 2004.

18. Kirby, *Animal Factory*, 25.

19. Wayne Pacelle, "Breaking News: Smithfield Will Complete Transition to Group Housing for Sows by End of 2017," *A Humane Nation*, Jan. 4, 2017, http://blog
.humanesociety.org/wayne/2017/01/smithfield-humane-housing-sows-2017.html.

20. Kimberlie Clyma, "Smithfield on Schedule for Gestations-Crate-Free Target," *Meat and Poultry*, Jan. 4, 2016, http://www.meatpoultry.com/articles/news_home
/Trends/2016/01/Smithfield_moves_closer_to_ges.aspx?ID= percent7B5CCFAC62
-D159-4081-B70E-0F193EB2CA06 percent7D&cck=1.

21. Joby Warrick and Pat Stith, "New Studies Show That Lagoons Are Leaking: Groundwater, Rivers Affected by Waste," *News & Observer*, Feb. 19, 1995, http://
www.pulitzer.org/winners/news-observer-raleigh-nc-work-melanie-sill-pat-stith
-and-joby-warrick.

22. "Key Industries: Hog Farming," *Learn NC*, http://www.learnnc.org/lp
/editions/nchist-recent/6257.

23. Warrick and Stith, "New Studies Show That Lagoons Are Leaking."

24. Christina Cooke, "North Carolina's Factory Farms Produce 15,000 Olympic Pools Worth of Waste Each Year," Civil Eats, June 28, 2016, http://civileats.com
/2016/06/28/north-carolinas-cafos-produce-15000-olympic-size-pools-worth-of
-waste/.

25. Nicolette Hahn Niman, foreword to *Righteous Porkchop: Finding a Life and Good Food beyond Factory Farms* (New York: Collins Living, 2009).

26. Ibid.

27. Environmental Working Group, "Exposing Fields of Filth: Locations of Concentrated Animal Feeding Operations in North Carolina by Watershed," ewg.org website, June 21, 2016, accessed Aug. 5, 2017, http://www.ewg.org/interactive-maps
/2016_north_carolina_animal_feeding_operations_bywatershed.php.

28. "North Carolina's Noxious Pig Farms," *New York Times*, Oct. 25, 2016, https://
www.nytimes.com/2016/10/25/opinion/north-carolinas-noxious-pig-farms.html.

29. Ibid.

30. Environmental Working Group, "Exposing Fields of Filth."

31. Christina Cooke, "North Carolina's Factory Farms Produce 15,000 Olympic Pools Worth of Waste Each Year," *CivilEats*, June 28, 2016, http://civileats.com
/2016/06/28/north-carolinas-cafos-produce-15000-olympic-size-pools-worth-of
-waste/.

32. Erica Hellerstein, "The N.C. Senate Overrides Cooper's HB 467 Veto: Hog-Farm-Protection Bill Is Law," *INDYWeek*, May 11, 2017, https://www.indyweek.com

/news/archives/2017/05/11/the-nc-senate-overrides-coopers-hb-467-veto-hog-farm
-protection-bill-is-law.

33. Diane Tennant and Linda McNatt, "Smithfield: Proud of the Company It
Keeps; Pollution Aside, Meatpacker Seen as Community Leader," *Virginian-Pilot*,
Jan. 29, 1997.

34. Donald P. Baker, "EPA Has Gone Hog Wild, VA Pork Baron Contends," *Washington Post*, Feb. 24, 1997.

35. Pat Stith and Joby Warrick, "Pork Barrels: Who's in Charge; Big Pork Pumps
Tens of Thousands of Dollars into North Carolina Political Campaigns," *News &
Observer*, Feb. 26, 1995.

36. Nicolette Hahn Niman, draft of foreword to *Righteous Porkchop*.

## Chapter 4

1. Henry Singleton, director of Family History–Genealogy, Bladenboro Historical
Society, interview by author, Bladenboro, NC, June 2016.

2. Sherri Buffkin, telephone interview by author, Aug. 13, 2011.

3. Keith Hempstead, "Plant Brings Life to Bladen, Officials Say," *Fayetteville
Observer*, Oct. 14, 1992, NewsBank record 504312.

4. Sherri Buffkin, telephone interview by author, March 6, 2011.

5. Descriptions of slaughtering process taken from NLRB transcript.

6. Howard Coble, "Favorite Breakfast Brains N' Eggs," *Congress Cooks*, accessed
Oct. 20, 2017, http://virtualcities.com/ons/nc/gov/ncgvhc1.htm.

7. NLRB transcript.

8. Ibid.

9. Ibid.

10. Sherri Buffkin, interview by author, Bladenboro, NC, Feb. 13, 2011.

11. Ludlum, interview by author, Bladenboro, NC, June 8–9, 2016.

12. NLRB transcript.

13. Rayshawn Ward, interview by author, Hope Mills, NC, June 8, 2016.

14. John Logan, "The Fine Art of Union Busting," *New Labor Forum* (Summer
2004): 77–85, 88–91, 147.

15. Nathan Shefferman, *The Man in the Middle* (Garden City, NY: Doubleday,
1961).

16. John Logan, "US Anti-Union Avoidance Consultants: A Threat to the Rights
of British Workers," Trades Union Congress, London, 2008, https://unionreps.org
.uk/sites/default/files/extras/loganreport.pdf.

17. NLRB transcript.

18. Jonathan Weisman and Ashley Parker, "Senate's Leader Sets Showdown over
Changes to Filibuster," *New York Times*, July 15, 2013, http://www.nytimes.com/2013
/07/16/us/politics/democrats-seeing-precedent-press-on-to-curb-filibuster.html;

Burgess Everett and John Bresnahan, "Obama Selects Labor Board Picks," *Politico*, July 16, 2013, http://www.politico.com/story/2013/07/white-house-consults-with -afl-cio-head-on-nlrb-picks-094280.

19. "Members of the NLRB since 1935," National Labor Relations Board, accessed Aug. 6, 2017, https://www.nlrb.gov/who-we-are/board/members-nlrb-1935.

20. Bill McMorris, "NLRB's Budget Doubles as Workload Plummets: Budget Up over Past 30 Years Despite Record-Low Union Membership," *Free Beacon*, Dec. 3, 2014, http://freebeacon.com/issues/nlrbs-budget-doubles-as-workload-plummets/.

21. Keith Ludlum, interview by author, June 8, 2016.

22. *Food, Inc.*, dir. Robert Kenner (Magnolia Pictures, Participant Media, and River Road Entertainment, 2008), mp4.

23. Gail A. Eisnitz, *Slaughterhouse: The Shocking Story of Greed, Neglect, and Inhumane Treatment inside the U.S. Meat Industry* (New York: Prometheus Books, 1997).

24. Ibid., 67.

25. Ibid., 265.

26. Ibid., 266.

27. Ibid., 84.

28. Ludlum, interview, June 8–9, 2016.

29. House Committee Reports, 110th Congress, H. Rept. 110-23—Employee Free Choice Act of 2007, https://www.congress.gov/congressional-report/110th-congress /house-report/23/1.

30. NLRB transcript.

31. Ibid.

## Chapter 5

1. NLRB transcript.

2. National Labor Relations Board, Respondent Smithfield Packing Company, Inc.'s Brief in Support of Its Exceptions to the Decision of the Administrative Law Judge, National Labor Relations Board, Region Eleven. In the Matter of: The Smithfield Packing Company, Inc.—Tar Heel Division and United Food and Commercial Workers Union Local 204, AFL-CIO, CLC. Case numbers 11-CA-15522, 15634, 15666, 15750, 15871, 15986, 16010, 16161, 16423, 16680, 17636, 17707, 17763, 17824, and 11-RC-6221 (hereafter "NLRB, Respondent"), Apr. 6, 2001.

3. Paul Woolverton, "Hog Plant to Phase Out Inmate Labor," *Fayetteville Observer*, Apr. 8, 1995, correction to headline: "The Plant Plans to Phase Out Only Those Prisoners Convicted of Violent Crimes," NewsBank record 440029.

4. NLRB transcript.

5. Ibid.

6. Ludlum, interview, June 8–9, 2016.

7. Ibid.

8. NLRB transcript.

9. Ibid.

10. Dave Jamieson, "This Is What It's Like to Sit through an Anti-Union Meeting at Work," *Huffington Post*, Sept. 3, 2004, http://www.huffingtonpost.com/2014/09/03/captive-audience-meetings-anti-union_n_5754330.html.

11. NLRB transcript.

12. Smithfield v. UFCW Local 204, NLRB Administrative Law Judge John H. West Decision #JD-158-00, Dec. 15, 2000. (Hereafter "NLRB, West Decision, 2000.")

13. NLRB transcript.

14. Ibid.

15. Ibid.

16. Ibid.

17. Ibid.

18. Jasper Brown, telephone interview by author, July 27, 2016.

19. Brown, telephone interview by author, July 28, 2016.

20. Ludlum, interview, June 8–9, 2016.

21. Ibid.

22. NLRB, West Decision, 2000.

23. Buffkin, interview, March 6, 2016.

24. Catherine Pritchard, "Workers, Plant Clash, Company Says Union Cost High, Organizers Disagree," *Fayetteville Observer*, Aug. 21, 1994, NewsBank record 457753.

25. Rogers Worthington, "Meatpacker Cudahy to Lay Off 70% of Workers," *Chicago Tribune*, Dec. 5, 1987, http://articles.chicagotribune.com/1987-12-05/business/8703310654_1_meatpacking-commercial-workers-meatcutters.

26. Della Pollock, *Remembering: Oral History Performance* (Palgrave Macmillan, 2005), 86.

27. Horowitz, *"Negro and White, Unite and Fight!,"* 270.

28. Hardy Green, *On Strike at Hormel: The Struggle for a Democratic Labor Movement* (Philadelphia: Temple University Press, 1990), 296.

29. Lance Compa, "A Second Look at the Hormel Strike," 1986, Cornell University ILR Collection, http://digitalcommons.ilr.cornell.edu/cgi/viewcontent.cgi?article=1081&context=workingpapers; David Moberg, foreword to *On Strike at Hormel: The Struggle for a Democratic Labor Movement* by Hardy Green (Philadelphia: Temple University Press, 1990), xii.

30. Horowitz, *"Negro and White, Unite and Fight!,"* 248.

31. Nelson Johnson, telephone interview by author, July 14, 2016.

32. NLRB transcript.

33. John Rene Rodriguez and Rayshawn Ward v. Smithfield Packing Co. Inc., et al., No. 5:00-CV-613-BR-2, Excerpts of Deposition of Rayshawn Ward, p. 44 (Ward Civil Suit).

34. "Smithfield Foods' Rotten Record," *Socialist Worker*, Dec. 15, 2006, http://socialistworker.org/2006-2/613/613_08_Smithfield.shtml.

35. Catherine Pritchard, "Workers, Plant Clash."

36. Horowitz, *"Negro and White, Unite and Fight!,"* 271.

37. Final Responsive Records Case No. 11-CA-15522. Document provided by NLRB in response to FOIA July 2016. Faxed document (appears to be a 5-page document) from Maupin Taylor, attorneys for Smithfield. FRI 10:46 (hole punch over a.m. or p.m.) March 16, 2001; transcript of anchor David Brancaccio and NPR reporter Leda Hartman interviewing Sherri Buffkin for "Marketplace," aired March 14, 2001.

38. Pritchard, "Workers, Plant Clash."

39. Ibid.

40. Scott Yates, "Workers Rally for Union," *Fayetteville Observer*, Aug. 21, 1994, NewsBank record 457758.

41. Ibid.

42. NLRB, West Decision, 2000.

43. Yates, "Workers Rally for Union."

44. Catherine Pritchard, "Food Union Rejected: Tar Heel Firm Wins Worker Vote," *Fayetteville Observer*, Aug. 26, 1994.

45. Ibid.

46. Catherine Pritchard, "Hog Company Denies Unfair Vote Influence, Organizer Vows to Keep Trying in Tar Heel," *Fayetteville Observer*, Aug. 27, 1994.

47. Ibid.

48. Ibid.

## Chapter 6

1. Keith Hempstead, "Plant Brings Life to Bladen, Officials Say," *Fayetteville Observer*, Oct. 14, 1992, NewsBank record 504312.

2. Paul Woolverton, "Bladen: Growing Economy Fills Coffers, Auditor Reports," *Fayetteville Observer*, Oct. 17, 1995, NewsBank record 424964.

3. Stella M. Hopkins, "Going Global," *Charlotte Observer*, Aug. 11, 1996, NewsBank record 960811067.

4. Ibid.

5. Ibid.

6. "Bladen County Hog Operation Boon and Bust," *Fayetteville Observer*, Dec. 4, 1995, NewsBank record 421083.

7. Eric Dyer, "Pork Producer to Close Lagoons: Smithfield Foods Ends Months of Negotiation with the States and Agrees to Spend $65 Million Testing Waste-Disposal Methods and Protecting the Environment," *News & Record* (Greensboro), July 25, 2000, http://www.greensboro.com/pork-producer-to-close-lagoons-smithfield-foods-ends-months-of/article_6b152d99-61ad-5449-9e28-3df9966429e4.html.

8. Josh Shaffer, "Town Supports Hog Plant," *Fayetteville Observer*, Jan. 24, 1998, NewsBank record 357072.

9. Michelle Washington, "State Considers Fining Hog Plant," *Fayetteville Observer*, Dec. 9, 1998, NewsBank record 16114.

10. Ibid.

11. Lorry Williams, "Smithfield Issued New Permit," *Fayetteville Observer*, May 11, 1999, NewsBank record 165324.

12. United States of America v. Smithfield Foods, Incorporated; Smithfield Packing Company, Incorporated, Gwaltney of Smithfield, LTD, Sept. 14, 1999, CA-96-1204-2, Fourth Circuit, http://www.ca4.uscourts.gov/Opinions/Published/972709.P.pdf.

13. Jim Spencer, "Rettig Not Only 'Rogue Employee' at Smithfield's," *Daily Press*, July 9, 1997, http://articles.dailypress.com/1997-07-09/news/9707090059_1_smith field-foods-water-quality-pagan-river.

14. Marl Weisbrot, Stephan Lefebvre, and Joseph Sammut, "Did NAFTA Help Mexico? An Assessment after Twenty Years," Center for Economic and Policy Research, Feb. 2014, http://cepr.net/documents/nafta-20-years-2014-02.pdf.

15. Laura Sanicola, "NAFTA's Effect on the Mexican Economy," *Fordham Political Review* (Aug. 10, 2015), http://fordhampoliticalreview.org/naftas-effect-on-the -mexican-economy/.

16. Rakesh Kochhar, Roberto Suro, and Sonya Tafoya, "The New Latino South: The Context and Consequences of Rapid Population Growth," Pew Research Center, *Hispanic Trends*, July 26, 2005, http://www.pewhispanic.org/2005/07/26/the-new -latino-south/.

17. North Carolina Budget and Management, County Hispanic Totals, April 2000 County Census Hispanic Populations with Growth and Migration from April 1990, https://www.osbm.nc.gov/county-hispanic-totals.

18. Ibid.

19. "Demographic Profile of Hispanics in North Carolina, 2014," Pew Research Center, *Hispanic Trends*, http://www.pewhispanic.org/states/state/nc/.

20. Jerry Kammer, "Immigration Raids at Smithfield: How an ICE Enforcement Action Boosted Union Organizing and the Employment of American Workers," Center for Immigration Studies, July 13, 2009, accessed Aug. 5, 2017, https://cis.org /Immigration-Raids-Smithfield-How-ICE-Enforcement-Action-Boosted-Union -Organizing-and-Employment.

21. Ana-María González Wahl, "Southern (Dis)Comfort?: Latino Population Growth, Economic Integration and Spatial Assimilation in North Carolina Micro-politan Areas," *Sociation Today* (Fall 2007), http://www.ncsociology.org/sociation today/ana.htm.

22. Robert Lamme, "Union Vote Starts Today in Tar Heel," *Fayetteville Observer*, Aug. 21, 1997, NewsBank record 369781; Robert Lamme, "Union's Efforts at Meat Plant May Hinge on Hispanic Workers," *Fayetteville Observer*, Aug. 15, 1997, News-Bank record 370338.

23. U.S. Department of Commerce, Economics, and Statistics Administration,

Bureau of the Census, 1990 Census of Population and Housing Unit Counts, North Carolina, https://www.census.gov/prod/cen1990/cph2/cph-2-35.pdf.

24. Sherri Buffkin testimony, U.S. Congress, Senate, Committee on Health Education, Labor, and Pensions, Workers' Freedom of Association: Obstacles to Forming a Union, 107th Cong., 2nd sess., June 20, 2002, Senate Hearing 107-5291. (Hereafter "Buffkin testimony.")

25. Ibid.

26. NLRB, West Decision, 2000.

27. Ibid.

28. Ibid.

29. Ibid.

30. Ibid.

31. Ibid.

## Chapter 7

1. NLRB transcript.

2. Buffkin, telephone interview by author, May 12, 2014.

3. Sherri Wright Buffkin v. The Smithfield Packing Company, Inc. and Jere Null, Case No. 7:00-CV-55-F (1) United States District Court for the Eastern District of North Carolina Southern Division. (Hereafter "Buffkin v. Smithfield and Null.")

4. Ibid. Jackson is deceased. Null has never publicly commented on the alleged encounter and did not return several phone messages requesting an interview for this book, specifically asking for comment on Buffkin's lawsuit that alleged harassment.

5. John Renee Rodriguez and Rayshawn Ward plaintiffs v. Smithfield Packing so. Inc. Et Al—No: 5:00-CV-613-BR-2 (hereafter "Rodriguez and Ward v. Smithfield"), p. 564.

6. Buffkin v. Smithfield and Null.

7. Null has not responded to requests for interviews to discuss the allegations. Smithfield Foods has declined to comment for the book through its public relations department.

8. Buffkin, interview, March 6, 2016.

9. Ibid.; Buffkin v. Smithfield and Null. *Beverly Hills, 90210* was an American drama series produced by Aaron Spelling that aired on Fox from 1990 to 2000. It centered on friendships, romantic relationships, and issues like date rape, gay rights, and substance abuse among a group of wealthy teens, a world away from Bladen County.

10. NLRB transcript.

11. Buffkin, interview, May 12, 2014.

12. Rodriguez and Ward v. Smithfield; NLRB transcript.

13. NLRB transcript.

14. Ibid.

15. Catherine Pritchard, "Jesse Jackson to Push Union at Hog Plant," *Fayetteville Observer*, Aug. 12, 1997, NewsBank record 370537.

16. Robert Lamme, "Carolina Food Executive Criticizes Union, Churches," *Fayetteville Observer*, Aug. 20, 1997, NewsBank record 369949.

17. Ibid.

18. Ibid.

19. Pritchard, "Jesse Jackson to Push Union at Hog Plant"; Robert Lamme, "Food Union Calls on Jackson to Unify Support," *Fayetteville Observer*, Aug. 15, 1997, NewsBank record 370343.

20. Ward, interview, June 8, 2016.

21. Ibid.

22. Ibid.

23. Ibid.

24. Ibid.

25. NLRB transcript.

26. Ward, interview, June 8, 2016.

27. NLRB transcript.

28. Ibid.

29. Robert Lamme, "Food Union Calls on Jackson to Unify Support," *Fayetteville Observer*, Aug. 15, 1997, NewsBank record 370343.

30. Robert Lamme, "Union Vote Starts Today in Tar Heel," *Fayetteville Observer*, Aug. 21, 1997, NewsBank record 369781.

31. Ibid.

32. Lamme, "Food Union Calls on Jackson to Unify Support."

33. Ibid.

34. Rodriguez and Ward v. Smithfield, Davis Dep. 79–81.

35. Lamme, "Food Union Calls on Jackson to Unify Support."

36. Buffkin, interview, March 6, 2016.

37. NLRB transcript.

38. Ibid.

39. Ibid.

40. Robert Lamme, "Union Is Rejected," *Fayetteville Observer*, Aug. 23, 1997, NewsBank record 369655.

41. NLRB transcript.

42. NLRB transcript.

## Chapter 8

1. The UFCW did not allow interviews with Chad Young or any other employees for this book, citing the "gag order" in the 2008 RICO settlement.

2. Rodriguez and Ward v. Smithfield.

3. Ibid.

4. Ibid.

5. Ibid.; NLRB transcript.

6. NLRB, West Decision, 2000.

7. NLRB transcript.

8. Ward, interview, June 8, 2016.

9. Ibid.

10. NLRB transcript.

11. Ibid.

12. Ward, interview, June 8, 2016.

13. NLRB transcript.

14. Ibid.; Rodriguez and Ward v. Smithfield.

15. Rodriguez and Ward v. Smithfield.

16. Ward, interview, June 8, 2016.

17. NLRB transcript.

18. Ibid.

19. Ibid.

20. Ibid.

21. Ibid.

22. Ibid.

23. NLRB transcript.

24. Ibid.

25. Ibid.

26. Buffkin, interview, March 6, 2016.

27. Ibid.

28. Ibid.

29. NLRB transcript.

30. Buffkin v. Smithfield and Null.

31. Ibid.

32. NLRB transcript.

33. Null has never publicly responded to Buffkin's claims.

34. Buffkin, interview, March 6, 2016.

35. Buffkin v. Smithfield and Null.

36. Ibid.

37. Buffkin, interview, Aug. 13, 2011.

38. Ibid.

39. Ibid.

40. Buffkin v. Smithfield and Null.

41. Buffkin testimony.

42. Brown, telephone interview by author, Feb. 16, 2016.

43. Brown, telephone interview by author, March 1, 2016.

44. Buffkin, interview, Aug. 13, 2011.

45. Buffkin testimony.

46. Donald Gattalaro, telephone interview by author, Feb. 17, 2016; Decision and Order by Chairman Battista and Members Liebman and Walsh, Dec. 16, 2004, affirming decision by Judge West in the Matter of: The Smithfield Packing Company, Inc. — Tar Heel Division and United Food and Commercial Workers Union Local 204, AFL-CIO, CLC. Case numbers 11-CA-15522, 15634, 15666, 15750, 15871, 15986, 16010, 16161, 16423, 16680, 17636, 17707, 17763, 17824, and 11-RC-6221. (Hereafter "Decision and Order.")

47. Brown, interview, Feb. 16, 2016.

## Chapter 9

1. Clif LeBlanc, "Hundreds Return Home to Columbia's Booker T. Washington," *The State*, June 15, 2013, http://www.thestate.com/news/local/civil-rights/article 14434580.html.

2. Willis J. Goldsmith of Jones Day law firm to Abby Simms, Esq. Office of the General Counsel, National Labor Relations Board, 18-page letter, March 31, 2005, Re: General Counsel's Investigation into the Conduct of William P. Barrett, Esq., Margie T. Case, Esq. and Joel H. Katz, Esq. in the matter of Smithfield Packing Co., 344 MLRB No. 1 (2004), p. 7.

3. NLRB, Respondent, Apr. 5, 2001, p. 4.

4. Gattalaro, interview, Feb. 17, 2016.

5. Julie Eisenberg, UFCW analyst, telephone interview by author, 2005 (exact date unknown; interview done prior to publication of "Slaughterhouse '05," *Portfolio Weekly*, July 5, 2005).

6. National Labor Relations Board, "NLRB Appoints Five Administrative Law Judges," press release, Document NLB2036 — 00042-NLB2036-000045, May 10, 2981, https://www.nlrb.gov/news-outreach/news-releases.

7. Gattalaro, interview, Feb. 17, 2016.

8. The case, which lasted from October 19, 1998, until July 19, 1999, was transcribed into 7,910 pages, eventually digitized into searchable documents provided by the NLRB for research on this book.

9. Buffkin, interview, March 6, 2016.

10. Brown, interview, Feb. 16, 2016.

11. Buffkin, interview, March 6, 2016.

12. Ibid.

13. NLRB transcript.

14. Larry Jackson did not take over the other responsibilities Buffkin had as production support manager, which included supervision of laundry, sanitation, and the warehouse.

15. NLRB transcript.

16. Ibid.

17. Ibid.

18. Ibid.

19. Susan Friend, "RADAR," *Daily Press*, Oct. 16, 1999.

20. David Ress, "No. 1 Richest Is Mars," *Daily Press*, July 1, 1992, http://articles
.dailypress.com/1992-07-01/business/9206300275_1_mars-family-daily-press
-hampton-roads.

21. Brown, interview, Feb. 16, 2016.

22. NLRB transcript.

23. NLRB transcript.


## Chapter 10

1. Goldsmith to Simms, p. 9.

2. NLRB, West Decision, 2000.

3. Gattalaro, interview, Feb. 17, 2016.

4. Buffkin, interview, Aug. 13, 2011.

5. Rodriguez and Ward v. Smithfield.

6. Ibid.

7. United Food and Commercial Workers Union (UFCW), "The Verdict Is In:
Smithfield's Use of Intimidation, Violence and False Arrests Violates Federal Civil
Rights Laws; Workers Win $755,000 in Jury Verdict Against Smithfield," *PR Newswire*
(dateline Raleigh, NC), press release, March 5, 2002.

8. United States Court of Appeals for the Fourth Circuit, John Renee Rodriguez;
Rayshawn Ward v. Smithfield Packing Company, Daniel M. Priest, et. al. CA-00-613
-BR (2), Argued: June 3, 2003. Decided: July 30, 2003.

9. Respondent's Exceptions to the Decision of the Administrative Law Judge. National Labor Relations Board, Region Eleven. In the Matter of: The Smithfield Packing Company, Inc.—Tar Heel Division and United Food and Commercial Workers
Union Local 204, AFL-CIO, CLC. Case numbers 11-CA-15522, 15634, 15666, 15750,
15871, 15986, 16010, 16161, 16423, 16680, 17636, 17707, 17763, 17824, and 11-RC-6221,
Apr. 6, 2001.

10. NLRB, Respondent, Apr. 6, 2001, p. 5.

11. Goldsmith to Simms.

12. NLRB, Respondent, April 6, 2001, p. 74.

13. Ibid.

14. Ibid., Apr. 5, 2001.

15. Protective Order, Sherri Wright Buffkin v. The Smithfield Packing Company,
Inc. and Jere Null, Case No. 7:00-CV-55-F (1) United States District Court for the
Eastern District of North Carolina Southern Division, July 12, 2000.

16. National Labor Relations Board, NLRB Motion in Opposition to Respondent's
Motion to Reopen the Record, National Labor Relations Board, Region Eleven. In the
Matter of: The Smithfield Packing Company, Inc.—Tar Heel Division and United
Food and Commercial Workers Union Local 204, AFL-CIO, CLC. Case numbers

11-CA-15522, 15634, 15666, 15750, 15871, 15986, 16010, 16161, 16423, 16680, 17636, 17707, 17763, 17824, and 11-RC-6221, Apr. 24, 2001.

17. Buffkin testimony.

18. Ibid.

19. Decision and Order.

20. Lynn Waltz, "A Collection of Reportage," MFA thesis, Old Dominion University, 2011.

21. Robertson, telephone interview by author, 2005 (exact date unknown; interview done prior to publication of "Slaughterhouse '05," *Portfolio Weekly*, July 5, 2005). Robertson worked at the time for Hunton & Williams, a Richmond, Virginia, law firm.

22. Ibid.

23. NLRB, West Decision, 2000.

24. Robertson, interview, 2005.

25. Ibid.

26. Tom Clarke, telephone interview by author, 2005 (exact date unknown; interview done prior to publication of "Slaughterhouse '05," *Portfolio Weekly*, July 5, 2005).

27. Tom Fredrickson, "Packer to Buy N.C. Hog Producer: $470 Million Deal Will Double Smithfield's Sows," *Daily Press*, Sept. 3, 1999, http://articles.dailypress.com /1999-09-03/business/9909070001_1_hog-producer-smithfield-s-shares-smithfield -chairman.

28. "The History of Smithfield Foods," *Virginian-Pilot*, May 29, 2013, http:// pilotonline.com/business/timeline-the-history-of-smithfield-foods/article_a2a34b25 -aeb2-5d29-ab26-54094bb62bfd.html.

29. Waltz, "Slaughterhouse '05."

30. Jim Papian, telephone interview by author, 2005 (exact date unknown; interview done prior to publication of "Slaughterhouse '05," *Portfolio Weekly*, July 5, 2005).

## Chapter 11

1. Arnold Ahlert, "Union Gangsters: John Sweeney, the Man Who Made the Union Gangster Way a Science," *FrontPage Mag*, Nov. 3, 2011, http://www.frontpagemag .com/fpm/111248/union-gangsters-john-sweeney-arnold-ahlert.

2. Fink and Alvis, *The Maya of Morganton*, 105.

3. Eduárdo Peña, interview by author, Red Springs, NC, Feb. 2005.

4. Eduardo Peña, interview by author, Feb. 2005.

5. Lance Compa, interview by author, Fayetteville, NC, Feb. 2005.

6. Lance Compa, *Union Time*, dir. Matthew Barr (Greensboro, NC: The Unheard Voices Project, 2016), film.

7. Terry Kilbride, various interviews by author, Fayetteville, NC, 2005.

8. Compa interview.

9. Ray Hall, interview by author, Fayetteville, NC, Feb. 2005.

10. QSI, Inc. and United Food and Commercial Workers Union; Smithfield Packing Company and United Food and Commercial Workers International Union. Cases 11-CA-20240, 11-CA-20317, 11-CA-20241, and 11-CA-20281. NLRB Decision and Order. Apr. 28, 2006. Note: Case heard on Apr. 11, 2005, by Administrative Law Judge Lawrence W. Cullen, appealed to NLRB board, then to U.S. Court of Appeals; United States Court of Appeals, Fourth Circuit. Smithfield Packing Company, Incorporated vs. National Labor Relations Board. Nos. 06-1541, 06-1652. Decided: Dec. 5, 2007.

11. Dan English, interviews by author, Robeson and Bladen Counties, NC, Feb. 2005.

12. Charlie LeDuff, "At the Slaughterhouse, Some Things Never Die: Who Kills, Who Cuts, Who Bosses Can Depend on Race," *New York Times*, June 16, 2000, https://partners.nytimes.com/library/national/race/061600leduff-meat.html.

13. "Kenneth McKinnon House, Robeson County," United States Department of the Interior, National Park Service, National Register of Historic Places, OMB No. 1024-0018, July 27, 2005, http://www.hpo.ncdcr.gov/nr/RB0520.pdf.

14. J. Kyle Foster, "2000 Yields Boom, Bust," *Fayetteville Observer*, Dec. 31, 2000, NewsBank record 671126.

15. J. Kyle Foster and Todd Leskanic, "Once-Mighty Mills in Decline," *Fayetteville Observer*, Oct. 15, 2000, NewsBank record 658875.

16. Venita Jenkins and April Johnston, "St. Pauls' Changing Face," *Fayetteville Observer*, July 15, 2007, NewsBank record 1162642.

17. Ibid.

18. Ibid.

19. Change to Win, "Change to Win Constitution (as amended November 2010), Preamble and Mission Statement," Strategic Organizing Center website, accessed Aug. 5, 2017, http://changetowin.org/wp-content/uploads/2017/06/CtW-Constitution -as-amended-2010-11.pdf.

## Chapter 12

1. Gene Bruskin interview, Feb. 18, 2017.

2. Julie Forster, "Who's Afraid of a Little Mud?," *Business Week*, May 21, 2001.

3. Gene Bruskin, *Union Time*, dir. Matthew Barr.

4. Congressional Record—Extension of Remarks, Hon. George Miller of California in the House of Representatives, Monday, Feb. 5, 2007, https://www.gpo.gov/fdsys /pkg/CREC-2007-02-05/pdf/CREC-2007-02-05-pt1-PgE260-3.pdf#page=1.

5. Ibid.

6. Bruskin, interview, March 9, 2016.

7. Gene Bruskin, telephone interview by author, Feb. 16, 2016.

8. NLRB transcript.

9. Bob Harbrant, "Comprehensive Campaigns," in *Union Power in the Future: A Union Activist's Agenda*, ed. Ken Gagala (Ithaca, NY: ILR Press, 1987).

10. Kate Bronfenbrenner and Robert Hickey, "Winning Is Possible: Successful Union Organizing in the United States—Clear Lessons, Too Few Examples," *Multinational Monitor* (June 2003), http://www.multinationalmonitor.org/mm2003 /062003/bronfenbrenner.html.

11. Bruskin, interview, Feb. 18, 2017.

12. Joe Crump, "The Pressure Is On: Organizing without the NLRB," *Labor Research Review*, vol. 1, no. 18, article 8 (1991), http://digitalcommons.ilr.cornell.edu/lrr /vol1/iss18/8.

13. Ibid.

14. Leila McDowell, telephone interview by author, March 10, 2016.

15. Leila McDowell's Facebook page, accessed March 10, 2016, https://www.face book.com/leila.mcdowell.

16. U.S. House of Representatives, "Hidden Tragedy"; "Meatpacking," United States Department of Labor, Occupational Safety and Health Administration, Safety and Health Topics, accessed Aug. 5, 2017, https://www.osha.gov/SLTC/meat packing/.

17. Research Associates of America, "Packaged with Abuse: Safety and Health Conditions at Smithfield Packing's Tar Heel Plant," Oct. 2006, revised and updated January 2007, accessed Oct. 24, 2017, http://meatiesohs.org/files/international _stories/smithfield_packaged_with_abuse.pdf.

18. Smithfield Foods, Inc. and Smithfield Packaging Company v. United Food and Commercial Workers International Union, et al., Civil Action No. 3:07cv641, United States District Court, E.D. Virginia, Richmond Division, Oct. 14, 2008. Hereafter "RICO."

19. RICO.

20. Ibid.

21. Leila McDowell, interview, March 10, 2016.

22. Bob Herbert, "Where the Hogs Come First," *New York Times*, June 15, 2006.

23. McDowell, interview, March 10, 2016.

24. Bennie Mitchell, telephone interview by author, 2005 (exact date unknown; interview done prior to publication of "Slaughterhouse '05," *Portfolio Weekly*, July 5, 2005).

25. Mitchell, interview, 2005.

26. Robert W. Cherny, "*The Jungle* and the Progressive Era," *History Now: The Journal of the Gilder Lehrman Institute of American History*, n.d., https://www.gilder lehrman.org/history-by-era/politics-reform/essays/jungle-and-progressive-era.

27. Bruskin, interview, March 9, 2016.

28. Gene Bruskin, telephone interview by author, Jan. 22, 2016.

29. Al Greenwood, "Region's Immigrants Join in 'National Cry,'" Apr. 11, 2006, NewsBank record 1075484.

30. Julia Oliver, "The Fight Hits Home," *Fayetteville Observer*, May 2, 2006, NewsBank record 1080058.

31. Ibid.

32. Ibid.

33. Ibid.

34. John H. West, telephone interview by author, March 11, 2016.

35. Compa, interview, 2005.

36. Katherine Torres, "Meatpackers Rally in Washington, D.C., to Push for Unionization, Safety," *EHS Today*, June 26, 2006, http://ehstoday.com/news/ehs_imp_38304. Luter IV did not respond to two phone messages left at his home in 2016 requesting comment.

37. Jeremiah McWilliams, "Joe Luter III to Step Down as Smithfield Foods' CEO," June 16, 2006, http://pilotonline.com/business/joe-luter-iii-to-step-down -as-smithfield-foods-ceo/article_7cc6405a-36a8-5106-b880-6c589dff9b08.html.

38. David Kesmodel, "Smithfield: How Sausage Was Made: Joe Luter Transformed His Family's Firm into a Global Giant," updated June 5, 2013, http://bambooinnovator .com/2013/06/06/smithfield-how-sausage-was-made-joseph-w-luter-iii-transformed -his-familys-firm-into-a-global-giant/.

39. Brown, interview, Feb. 16, 2016.

40. Ludlum, interview, June 8–9, 2016.

41. Sherri Buffkin, of course, was not there to celebrate. As part of the anti-union management, she was never part of the NLRB case, except to testify.

42. Ludlum, interview, June 8–9, 2016.

43. Ludlum, interview, June 8–9, 2016.

44. NLRB, West Decision, 2000.

45. Ibid.

46. House Hearing before the Subcommittee on Health, Employment, Labor and Pensions, U.S. House of Representatives, 110th Congress, First Session, Washington D.C., Feb. 8, 2007, https://www.gpo.gov/fdsys/pkg/CHRG-110hhrg32906/html /CHRG-110hhrg32906.htm (hereafter "House Hearing").

47. Bruskin interview, Jan. 22, 2016.

## Chapter 13

1. Gene Bruskin, telephone interview by author, Feb. 6, 2016.

2. Johnson, interview, July 14, 2016.

3. Donald W. Patterson, *News & Record* (Greensboro, NC), July 10, 2006, NewsBank record 060712988375.

4. Julia Oliver, "Packer's Practices Decried," *Fayetteville Observer*, July 22, 2006, NewsBank record 1100574.

5. Amanda Greene, "Objects to Stores' Buying from Tar Heel Pork Plant," *StarNews* (Wilmington, NC), Dec. 3, 2006, NewsBank record 173314.

6. RICO.

7. Ibid.

8. Ibid.

9. "Frederick Morganthall, Divisional Executive VP at Kroger," FindTheCompany by GRAPHIQ, accessed Aug. 5, 2017, http://executives.findthecompany.com/l/78295 /Frederick-J-Morganthall-II.

10. "7625 Stonecroft Park Dr.," Zillow, Inc., accessed March 12, 2017, https://www .zillow.com/homedetails/7625-Stonecroft-Park-Dr-Charlotte-NC-28226/65010247 _zpid.

11. Bruskin, interview, March 9, 2016.

12. RICO.

13. Ibid.

14. Ibid.

15. Gail D. McAfree, "Workers at Smithfield Plant Are Human," *Fayetteville Observer*, Apr. 3, 2007, NewsBank record 1147343.

16. Walt Taylor, *Bladen Journal*, July 20, 2007, NewsBank record bdj/raw/articles /2007/07/20/news/community/news01.txt.

17. Staff report, *Fayetteville Observer*, July 31, 2007, NewsBank record 1164634.

18. John Fuquay, "NAACP Issues Boycott Threat," *Fayetteville Observer*, Aug. 3, 2007, Newsbank record 1165040.

19. RICO.

20. Ibid.

21. Ibid.

22. Venita Jenkins, "Image Battle Clouds Issues," *Fayetteville Observer*, July 29, 2007, NewsBank record 1164398.

23. Dave Mayfield, "Smithfield Foods Drops Paula Deen as Spokeswoman," *Virginian-Pilot*, June 24, 2013.

24. Bruskin, interview, Feb. 16, 2016.

25. Bruskin, interview, Feb. 6, 2016.

26. Ibid.

27. RICO.

28. Marian Burros, "Union, in Organizing Fight, Tangles with Celebrity Cook," *New York Times*, Apr. 20, 2007, https://mobile.nytimes.com/2007/04/20/us/20deen.html.

29. Kirsten Valle, "60 Demonstrate at Deen Visit—They Want Her to Stop Endorsing Pork Firm That's in Union Dispute," *Charlotte Observer*, July 1, 2007, NewsBank record 0707010136.

30. "Interview with Paula Deen," *Larry King Live*, CNN, aired Aug. 6, 2007, http://transcripts.cnn.com/TRANSCRIPTS/0708/06/lkl.01.html.

31. "*Christmas with Paula Deen* (Simon & Schuster)," *The Diane Rehm Show*, WAMU 88.5, distributed by NPR, aired Nov. 28, 2007, http://dianerehm.org/?s =Paula+Deen.

32. McDowell, interview, March 10, 2016.

33. "*Christmas with Paula Deen* (Simon & Schuster)."

34. Smithfield Foods, Inc., and Smithfield Packing Company v. United Food and Commercial Workers International Union, United States District Court for the Eastern District of Virginia, Richmond Division, Civil Action No. 3:07CV641, Memorandum in Opposition to Defendants' Motion to Dismiss, Dec. 21, 2007, http://graphics8 .nytimes.com/packages/pdf/national/smithfield_opposition.pdf.

35. Peter Hamby and Suzanne Malveaux, "Thanks to Oprah, Obama Camp Claims Biggest Crowd Yet," CNNPolitics, Dec. 9, 2007, http://www.cnn.com/2007 /POLITICS/12/09/oprah.obama.

36. Fred Lucas, "Obama Administration Has Given Obamacare Waivers to 28 Food Workers Union Locals—Union's PAC Spent $673,309 to Get Obama Elected," CNS News, Jan. 31, 2011, http://www.cnsnews.com/news/article/obama-administration -has-given-obamacare-waivers-28-food-workers-union-locals-union-s.

37. Nick Baumann, "The SEIU Picks Obama," *Mother Jones*, Feb. 14, 2008, http:// www.motherjones.com/mojo/2008/02/seiu-picks-obama.

## Chapter 14

1. Ludlum, interview, June 8–9, 2016.

2. Keith Ludlum, *Union Time*, dir. Matthew Barr.

3. Chris Brooks, "The Volkswagen Defeat Wasn't Inevitable—and Labor Can Still Win in the South," *In These Times*, Feb. 14, 2017, http://inthesetimes.com/working /entry/19898/the_volkswagen_.

4. Ludlum, interview, June 8–9, 2016.

5. Ibid.

6. David Bacon, "Unions Come to Smithfield," *American Prospect*, Dec. 17, 2008.

7. David Bacon, *Illegal People: How Globalization Creates Migration and Criminal Immigrants* (Boston: Beacon Press, 2009), 17.

8. Unnamed source, telephone interview by author, 2016.

9. Bacon, *Illegal People*, 18.

10. Claire Parker, "Smithfield Workers Protest Screenings," *Fayetteville Observer*, Nov. 17, 2006, NewsBank record 1124973.

11. Carlos Arce, telephone interview by author, March 11, 2016.

12. Parker, "Smithfield Workers Protest Screenings."

13. Ibid.

14. Arce, interview, March 11, 2016.

15. Unnamed source, interview, 2016.

16. Al Greenwood, "Smithfield Workers Return," *Fayetteville Observer*, Nov. 19, 2006, NewsBank record 1125347.

17. Arce, interview March 11, 2016.

18. Ibid.

19. Unnamed source, interview, 2016.

20. Jennifer Plotnick, "Workers Want Holiday," *Fayetteville Observer*, Jan. 11, 2007, NewsBank record 1133534.

21. Bruskin, interview, Feb. 21, 2016.

22. Sue Stock, "Laborers Stage MLK Day Protest," *News & Observer* (Raleigh, NC), Jan. 16, 2007, NewsBank record jbya6789.

23. Jerry Kammer, "The 2006 Swift Raids: Assessing the Impact of Immigration Enforcement Actions at Six Facilities," Center for Immigration Studies, March 2009, http://cis.org/2006SwiftRaids.

24. Jennifer Plotnick, "Plant Workers Arrested," *Fayetteville Observer*, Jan. 25, 2007, NewsBank record 1135476.

25. Bacon, *Illegal People*, 13.

26. Bruskin, interview, Jan. 22, 2016.

27. Editorial, *Winston-Salem Journal*, Feb. 8, 2007, NewsBank record 0702080044.

28. Bacon, *Illegal People*, 13.

29. Jennifer Plotnick, "Scared Workers Stay Home," *Fayetteville Observer*, Jan. 26, 2007.

30. Arce, interview, March 11, 2016.

31. Venita Jenkins and Jennifer Plotnick, "Fear of Deportation," *Fayetteville Observer*, Feb. 4, 2007.

32. Ibid.

33. Kevin Maurer, "Latinos Walk Off the Job," *Fayetteville Observer*, Jan. 28, 2007, NewsBank record 1136114.

34. Al Greenwood, "Policy Trims Plant's Staff," *Fayetteville Observer*, Feb. 22, 2007, NewsBank record 1139845.

35. Bacon, *Illegal People*, 19.

36. Ibid.

37. Venita Jenkins, "Families Tell of Arrest Fears," *Fayetteville Observer*, Feb. 8, 2008, NewsBank record 1137745.

38. Ibid.

39. "Illegal Immigrants" (editorial), *Winston-Salem Journal*, Feb. 8, 2007, NewsBank record 0702080044.

40. Venita Jenkins, "Immigration Agents Detain 28 in Raids," *Fayetteville Observer*, Aug. 23, 2007, NewsBank record 1167907.

41. Peña, interview, Feb. 14, 2016.

42. Ludlum, interview, June 8–9, 2016.

43. Peña, interview, Feb. 14, 2016.

44. Ludlum, interview, June 8–9, 2016.

45. *Food, Inc.*, dir. Robert Kenner.

46. Ibid.

47. Ibid.

48. Venita Jenkins, "14 Enter Immigration Pleas," *Fayetteville Observer*, Nov. 24, 2007, NewsBank record 1180403.

49. Paul Woolverton, "Detainees Appear in Federal Court," *Fayetteville Observer*, Aug. 29, 2007, NewsBank record 1168566.

50. Venita Jenkins, "14 Enter Immigration Pleas."

51. Arce, interview, March 11, 2016.

## Chapter 15

1. Steven Greenhouse, "Labor Board's Exiting Leader Responds to Critics," *New York Times*, Aug. 29, 2011, http://www.nytimes.com/2011/08/30/business/national -labor-boards-leader-leaves-amid-criticism.html.

2. Wilma Liebman, "Decline and Disenchantment: Reflections on the Aging of the National Labor Relations Board," *Berkeley Journal of Employment and Labor Law*, vol. 28, no. 2 (2007): 569–589, http://www.jstor.org/stable/24052247.

3. Ibid.

4. Ibid.

5. Ibid.

6. David Cay Johnston, "Income Gap Is Widening, Data Shows," *New York Times*, March 29, 2007.

7. Liebman, "Decline and Disenchantment."

8. House Hearing.

9. Ludlum, interview, June 8–9, 2016.

10. Ibid.

11. Ibid.

12. Ludlum, interview, June 8–9, 2016.

13. House Hearing.

14. Ludlum, interview, June 8–9, 2016.

15. "Employee Free Choice Act of 2007—Motion to Proceed," *Congressional Record*, vol. 153, no. 4, Senate GPO, June 26, 2007, S8378–S8398, https://www.gpo.gov /fdsys/pkg/CREC-2007-06-26/html/CREC-2007-06-26-pt1-PgS8378-2.htm.

16. Ludlum, interview, June 8–9, 2016.

17. Ibid.

18. Ibid.

19. Gene Bruskin, "If We Can Change the White House, We Can Change the Hog House," *New Labor Forum* (Dec. 31, 2010), http://newlaborforum.cuny.edu/2010/12 /30/if-we-can-change-the-white-house-we-can-change-the-hog-house/#sthash .LAqHdA2b.dpuf.

20. Ludlum, interview, June 8–9, 2016.

21. Ibid.

22. RICO.

23. Ibid.

24. Ibid.

25. Gene Bruskin, interview by author, Feb. 18, 2017.

26. Gene Bruskin, telephone interview by author, Feb. 21, 2016.

27. Venita Jenkins, "Workers Protest for Union," *Fayetteville Observer*, Aug. 29, 2007, NewsBank record 1168631.

28. *Food, Inc.*, dir. Robert Kenner.

29. Nelson Johnson, *Union Time*, dir. Matthew Barr.

30. Venita Jenkins, "Shareholders Get an Earful," *Fayetteville Observer*, Aug. 30, 1997, NewsBank record 1168831.

31. William J. Barber II, *Union Time*, dir. Matthew Barr.

32. William J. Barber II, *Food, Inc.*, dir. Robert Kenner.

33. Venita Jenkins, "Shareholders Get an Earful."

34. Bruskin, interview, Feb. 21, 2016.

35. Ibid.

36. Ibid.

37. Ludlum, interview, June 8–9, 2016.

38. Venita Jenkins, "Shareholders Get an Earful."

39. Terry Slaughter, *Union Time*, dir. Matthew Barr.

40. Venita Jenkins, "Shareholders Get an Earful."

41. Nelson Johnson, *Union Time*, dir. Matthew Barr.

42. Bruskin, interview, Feb. 21, 2016.

43. Venita Jenkins, "Smithfield Talks End," *Fayetteville Observer*, Oct. 16, 2007, NewsBank record 1175171.

44. *NOW*, show 250, Dec. 15, 2006, PBS, http://www.pbs.org/now/transcript/250.html.

45. Bruskin, interview, Feb. 18, 2017.

## Chapter 16

1. Bruskin, interview, Feb. 18, 2017.

2. Bruskin, interview, Jan. 22, 2016.

3. Julius G. Getman, *Restoring the Power of Unions: It Takes a Movement* (New Haven, CT: Yale University Press, 2010), Kindle edition, location 4958.

4. RICO.

5. Ibid.

6. Bruskin, interview, Feb. 18, 2017.

7. Adam Liptak, "A Corporate View of Mafia Tactics: Protesting, Lobbying and Citing Upton Sinclair," *New York Times*, Feb. 5, 2008, http://www.nytimes.com/2008/02/05/us/05bar.html.

8. Ibid.

9. Getman, *Restoring the Power of Unions*, location 4958.

10. Chris Flores, "Smithfield Foods: Union's Tactics Questioned," *Daily Press*, Oct. 21, 2007.

11. Liptak, "A Corporate View of Mafia Tactics."

12. Ibid.

13. Ibid.

14. Smithfield Foods, Inc. and Smithfield Packing Company, Plaintiffs, v. United Food and Commercial Workers International Union, et al., Defendants. Civil Action No. 3:07cv641. United States District Court for the Eastern District of Virginia, Richmond Division. May 29, 2008. (Hereafter "Smithfield Foods, Inc. and Smithfield Packing Company, Plaintiffs, v. United Food and Commercial Workers International Union et al.")

15. Ibid.

16. Getman, *Restoring the Power of Unions*, location 5033.

17. Ibid., location 5049.

18. Bruskin, interview, Jan. 22, 2016.

19. Getman, *Restoring the Power of Unions*, location 5081.

20. Bruskin, interview, Jan. 22, 2016.

21. Ibid.

22. Smithfield Foods, Inc. and Smithfield Packing Company, Plaintiffs, v. United Food and Commercial Workers International Union et al.; United States District Court for the Eastern District of Virginia, Richmond Division. Memorandum Opinion, Robert E. Payne, Senior District Judge, Oct. 23, 2008, http://www.leagle.com /decision/20081400585byfsupp2d815_11324/SMITHFIELD%20FOODS%20v.%20 UNITED%20FOOD%20AND%20COMM.%20WORKERS.

23. Bruskin, interview, Jan. 22, 2016.

24. Barack Obama, *Union Time*, dir. Matthew Barr.

25. Don Worthington, "1-on-1 with Obama," *Fayetteville Observer*, Apr. 18, 2008, NewsBank record 1201898.

26. Bruskin, interview, Jan. 22, 2016.

27. Ibid.

28. Ben James, "Smithfield Settles RICO Suit over Union Campaign," *Law360*, Oct. 26, 2008, https://www.law360.com/articles/74251/smithfield-settles-rico-suit -over-union-campaign.

29. Bruskin, interview, Feb. 21, 2016.

30. Unnamed source, interview, 2016.

31. Jeffrey S. Passel and D'Vera Cohn, "Mexican Immigrants: How Many Come? How Many Leave?," Pew Research Center, Hispanic Trends, July 22, 2009, accessed Aug. 5, 2017, http://www.pewhispanic.org/2009/07/22/mexican-immigrants-how -many-come-how-many-leave/.

32. Jerry Kammer, "Immigration Raids at Smithfield: How an ICE Enforcement Action Boosted Union Organizing and the Employment of American Workers," Center for Immigration Studies, July 13, 2009, accessed Aug. 5, 2017, https://cis.org /Immigration-Raids-Smithfield-How-ICE-Enforcement-Action-Boosted-Union -Organizing-and-Employment.

33. John Ramsey and Sarah Reid, "Race and the Union," *Fayetteville Observer*, Dec. 21, 2008, NewsBank record 1235960.

34. Ludlum, interview, June 8–9, 2016.

35. Bruskin, interview, Feb. 18, 2017.

36. Ludlum, interview, June 8–9, 2016.

37. Unnamed source, interview, 2016.

38. *Union Time*, dir. Matthew Barr.

39. Unnamed source, interview, 2016.

40. Bacon, "Unions Come to Smithfield."

41. Eduardo Peña, *Food, Inc.*, dir. Robert Kenner.

42. Kristin Collins, "Raids May Have Aided Smithfield Union Vote," *Charlotte Observer*, Jan. 3, 2009.

43. Ernsberger, "The Ham Man."

44. Gene Bruskin, *Union Time*, dir. Matthew Barr.

45. "United Food & Commercial Workers, Local 1208," Union Facts, accessed Aug. 5, 2017, https://www.unionfacts.com/lu/544130/UFCW/1208/.

46. Ludlum, interview, June 8–9, 2016.

## Epilogue

1. "Bankruptcies," *Fayetteville Observer*, March 27, 2000, NewsBank record 418890.

2. Buffkin, interview, Feb. 13, 2011.

3. Ludlum, interview, June 8–9, 2016.

4. Ibid.

5. The "1 percent" refers to the wealthiest 1 percent of the population in the United States.

6. Keith Ludlum, "Keith Ludlum: Community Support Is Key #Organizethe South," YouTube, "Organize the South," North Carolina State AFL-CIO, panel at Duke University, published on Feb. 28, 2014, accessed Aug. 5, 2017, https://www.youtube.com/watch?v=lpPLbTUKci8.

7. Michael Futch, "Actor, Activist Danny Glover Rallies for Union Members at Mountaire Farms," *Fayetteville Observer*, Aug. 6, 2014, accessed March 12, 2017, http://www.fayobserver.com/news/local/actor-activist-danny-glover-rallies-for-union-members-at-mountaire/article_2ab7baf6-5644-5b8e-8ca2-790f0c5287c2.html.

8. Ludlum, interview, June 8–9, 2016.

9. Bruskin, interview, Feb. 18, 2017.

10. Buffkin, interview, Aug. 13, 2011.

11. Larry Johnson, interview by author, Elizabethtown, NC, June 9, 2016.

12. Emily Boster, "Prestage Farms Chooses North Iowa for Pork Processing Plant," KIMT.com, March 26, 2016, accessed March 12, 2017, http://kimt.com/2016/03/21/prestage-foods-pick-north-iowa/.

13. MyKayla Hilgart, "Prestage Protest Draws Crowds Monday Morning," KIMT.com, Apr. 25, 2016, accessed March 12, 2017, http://kimt.com/2016/04/25/prestage-protest-planned-monday-morning/.

14. "The People vs. Prestage in Mason City," Facebook page, accessed Aug. 6, 2017, https://www.facebook.com/search/top/?q=%E2%80%9CThe%20People%20vs.%20 Prestage%20in%20Mason%20City%E2%80%9D%20.

15. Donnelle Eller, "Analyst: New Pork Processing Facilities Threaten Older Iowa Plants," *Des Moines Register*, June 10, 2016, http://www.desmoinesregister.com/story /money/agriculture/2016/06/10/analyst-new-pork-processing-facilities-threaten-older -iowa-plants/85700296/; Matt Bradley, "Prestage Interview: Racism Alive and Well, Opposition Group 'Kooks,'" KIMT.com, May 5, 2016, http://kimt.com/2016/05/05 /prestage-interview-racism-alive-and-well-opposition-group-kooks/.

16. Brian Tabick, "Wright County Officials Outline Limits to Prestage Hearing," KIMT.com, July 18, 2016, http://kimt.com/2016/07/18/wright-county-board-of -supervisors-vote-to-rezone-land-for-proposed-pork-processing-plant/.

17. "Musings on Stylish Living Dinner in the Country at the Pine Creek Sporting Club Home of Karin and Joe Luter," Peak of Chic website, Sept. 4, 2012, accessed Aug. 6, 2017, http://thepeakofchic.blogspot.com/2012/09/palm-beach-entertaining. html; http://www.patrickmcmullan.com/users/login.aspx?url=site percent2fsearch .aspx percent3ft percent3dperson percent26s percent3dKarin percent2520Luter percent26page percent3d2 percent26pgSize percent3d64 percent26sortdir percent3d DESC, accessed March 12, 2017; Property of Joseph and Karin Luter, Palm Beach County Property Appraiser, Dorothy Jacks, CFA, accessed Aug. 6, 2017, http://pbcgov .com/papa/Asps/PropertyDetail/PropertyDetail.aspx?parcel=50434234030000461 &srchtype=ADV&owner=Luter&streetno=&prefix=&streetname=&suffix=&postdir =&unitno=&srchparcel=&range=&twp=&section=&book=&page=&legal=&subdiv =&muni=&zip=&usetype=&condo=; "'A Garden in the Wild' Dinner Dance Raises $1.19 Million," Palm Beach Zoo & Conservation Society, Feb. 4, 2014, accessed March 12, 2017, http://www.palmbeachzoo.org/garden-in-the-wild-dinner-dance-raises -one-million-for-palm-beach-zoo?returnTo=main.

18. Bruskin, interview, Jan. 22, 2016.

19. Ludlum, interview, June 8–9, 2016.

# BIBLIOGRAPHIC ESSAY

Most of the primary research for this book—interviews and analysis of court records—was to serve the narrative of how the world's largest slaughterhouse was unionized. Additional research was done to provide context for that story. One primary document, provided by the NLRB, was the 7,910-page transcript of the NLRB hearing of hundreds of charges against Smithfield Packing held over nine months between 1998 and 1999. The transcript included the testimony of about 130 witnesses and arguments from NLRB attorneys and attorneys representing Smithfield and the UFCW. Another document, readily available online, was the 436-page ruling of Judge John H. West that resulted from that hearing. Additional information came from two civil lawsuits filed in North Carolina state courts, one by Sherri Buffkin and the other by worker Rayshawn Ward and union organizer John Rodriguez. Additional documents, provided by the NLRB and relating to Smithfield's appeals to its rulings, are not available to the public. I also relied on documents filed in Richmond federal court when Smithfield Foods sued the UFCW and multiple other parties.

Other primary research included interviews with the following individuals either in person or on the phone: Carlos Arce (Father Carlos), then priest of St. Andrews Catholic Church; Jasper Brown, former NLRB attorney arguing charges against Smithfield Packing; Gene Bruskin, former director of the UFCW union campaign at the Tar Heel plant; Sherri Buffkin, former production support manager at Smithfield Packing; Tom Clarke, then UFCW organizer; Lance Compa, then labor studies professor at Cornell University; Julie Eisenberg, former UFCW analyst; Dan English, then UFCW organizer; Donald Gattalaro, former NLRB attorney; Ray Hall, former employee at Smithfield Packing; Larry Johnson, former plant manager at Smithfield Packing; Nelson Johnson, co-founder of the Southern Faith, Labor and Community Alliance in Greensboro; Robert F. Kennedy Jr., environmental activist; Terry Kilbride, then workers' compensation attorney; Wilma Liebman, former chair, NLRB; Keith Ludlum; illegally fired former worker at Smithfield Packing and former president of UFCW Local 400; Joseph Luter IV, then executive vice president of Smithfield Foods; Leila McDowell, former head of communications for the UFCW union campaign at the Tar Heel plant; Bennie Mitchell, then chair of labor relations for the National Baptist Convention; Jim Papian, then UFCW spokesperson; Eduardo Peña, then UFCW organizer; Larry Pope, then president and COO of Smithfield Foods; Greg Robertson, labor attorney for Smithfield Packing; Henry Singleton, director of Family History–Genealogy, Bladenboro Historical Society; Rayshawn Ward, illegally fired former worker at Smithfield Packing; Judge John H. West, retired administrative law judge, NLRB.

In addition, I relied on hundreds of articles written for the *Fayetteville Observer*, which covered Smithfield Packing extensively through the years, as well as Charlie LeDuff's brilliant *New York Times* piece "At a Slaughterhouse, Some Things Never Die" and the Pulitzer Prize–winning series "Boss Hog," published by the Raleigh *News & Observer*. While I was writing this book, Matthew Barr, of the University of North Carolina at Greensboro, finished his documentary *Union Time*, which provides live action coverage of some events I had not attended.

These sources provided information specifically about the events at the Smithfield Foods plant, but to be fully appreciated the story needs to be placed within a historical, sociological, and industrial context, including the New Deal and federal labor law; the history of meatpacking and meatpacking unions; immigration and its impact on meatpacking; modern meatpacking practices; factory farms and their impact on the environment; and the history of slavery and workers in southeastern North Carolina, among others.

In terms of shedding light on the ethical considerations behind our individual food choices, two best-selling books led the way. Eric Schlosser's *Fast Food Nation: The Dark Side of the All-American Meal* (Houghton Mifflin, 2001) and Michael Pollan's *The Omnivore's Dilemma: A Natural History of Four Meals* (Penguin, 2006) were groundbreaking in terms of inspiring American self-reflection on the high cost of getting food to the table ethically. Schlosser, like some modern-day Upton Sinclair, exposed the dangers of meatpacking to a wide audience, vividly describing workers racing against line speed in a cattle slaughterhouse. Pollan took a step back from Schlosser's visceral style with his rational, deeply reflective exploration of how we produce the food we eat.

To better understand the role of Smithfield Foods in the larger modern agribusiness transformation of the latter half of the twentieth century, I turned to a series of outstanding books on modern meatpacking. *The Meat Racket: The Secret Takeover of America's Food Business* (Simon & Schuster, 2014), by Christopher Leonard, a former reporter for the Associated Press, examined the way a small number of corporations took over the nation's meat supply and in the process exploited farmers, consumers, and the environment while locking down their monopolies. Wilson J. Warren's *Tied to the Great Packing Machine: The Midwest and Meatpacking* (University of Iowa Press, 2007), is a topical exploration of meatpacking in the Midwest, including the role of women, race, workers' rights, humane treatment of animals, and the slaughtering process.

Meanwhile, several books helped me better understand the history of industrial farming in this country. *Animal Factory: The Looming Threat of Industrial Pig, Dairy, and Poultry Farms to Humans and the Environment* (St. Martin's, 2010), by David Kirby, fastidiously examines the high cost of industrial factory farming and the ignorance of many Americans regarding where their food comes from, specifically exploring confined animal feeding operations, or CAFOS. Kirby focuses on the damaging hog lagoons in North Carolina, spending time with environmentalist Rick Dove, who has been prophetic in his warnings of ecological devastation from the billions of gallons of hog waste sitting untreated in the state.

Kirby's book dovetails nicely with Nicolette Hahn Niman's *Righteous Porkchop: Finding a Life and Good Food beyond Factory Farms* (Collins Living, 2009), which chronicles the author's work as an environmental lawyer for the Waterkeeper Alliance, exposing the practices of hog factory farms, animal confinement, and water and air pollution. *The Chain: Farm, Factory, and the Fate of Our Food* (Harper, 2014), by Ted Genoways, takes the reader into the lives of slaughterhouse workers, including Latino immigrants, while telling the broad story of the industrialization of meatpacking. Vanesa Ribas's *On the Line: Slaughterhouse Lives and the Making of the New South* (University of California Press, 2016), meanwhile, is a first-person examination of factory-worker life in a small Smithfield Foods slaughterhouse in North Carolina. Ribas spent sixteen months from 2010 to 2012, working elbow to elbow on the slaughterhouse line with mostly Latino/a and African American workers to give an insider's view, with colorful, emotional descriptions. Similarly, David Bacon, who has written extensively about immigration in meatpacking and taken poignant photographs of workers, provides insight into the broader picture with *Illegal People: How Globalization Creates Migration and Criminal Immigrants* (Beacon Press, 2009). Another powerful piece of investigative work is *Slaughterhouse: The Shocking Story of Greed, Neglect, and Inhumane Treatment Inside the U.S. Meat Industry*, by Gail Eisnitz (Prometheus Books, 2007), an excruciating account of the horrors inflicted on animals in modern meatpacking based on interviews with workers. *Slaughterhouse Blues: The Meat and Poultry Industry in North America*, by Donald D. Stull and Michael J. Broadway (2nd ed.; Wadsworth, Cengage Learning, 2013), explores the impact of beef-, pork-, and poultry-processing plants on workers and communities throughout North America, together with the impact of industrial farming, which has decimated the small farmers in North Carolina.

Regarding modern labor law, two books in particular stand out: *Franklin D. Roosevelt and the New Deal: 1932–1940*, by William E. Leuchtenburg (1963; Harper Perennial, 2009), and *Senator Robert F. Wagner and the Rise of Urban Liberalism*, by J. Joseph Huthmacher (Atheneum, 1968). The influence of Robert Wagner (the National Labor Relations Act of 1935 was unofficially called the "Wagner Act") on modern society was both surprising and inspirational as a foundation for this book. In turn, *Broken Promises: The Subversion of U.S. Labor Relations Policy, 1947–1994*, by James A. Gross (Temple University Press, 1995), gives a good overview of the erosion of those worker rights.

Key writers on the narrower subject of meatpacking unions are historians Roger Horowitz and Rick Halpern. Particularly helpful to me was Horowitz's *"Negro and White, Unite and Fight!": A Social History of Industrial Unionism in Meatpacking, 1930–90* (University of Illinois Press, 1997), which traces the role of African Americans in the packinghouse and their struggle for collective bargaining with unions that were as segregated as the workplaces they sought to represent. Halpern's *Down on the Killing Floor: Black and White Workers in Chicago's Packinghouses, 1904–54* (University of Illinois Press, 1997), offers a detailed look at race relations and the formation of unions in the first half of the century.

Because the UFCW declined to comment for the book, citing a federal judge's ruling as essentially a gag order, it was difficult to gauge exactly what was going on with international union leadership before, during, and after the Smithfield Packing union campaign. I gained insight from Della Pollock's edited volume *Remembering: Oral History Performance* (Palgrave Macmillan, 2005), which gives voice to the workers of the UFCW Local P-40 at the Patrick Cudahy meatpacking plant in Cudahy, Wisconsin, who went on strike shortly after Smithfield Foods purchased the factory in 1984. Similarly, *On Strike at Hormel: The Struggle for a Democratic Labor Movement* (Temple University Press, 1990), by Hardy Green, gives an insider's view of the bitter internal feud between the international officers of the UFCW and those at the UFCW local during the violent strikes against Hormel in Austin, Minnesota, in 1985 and 1986, one of the most notorious union fights of the latter twentieth century, which was resolved shortly before Smithfield Foods opened the world's largest slaughterhouse in Tar Heel, North Carolina, in 1992.

# INDEX

*Page numbers in italics refer to photos and figures.*

"A people united . . ." chant, 160
absenteeism and turnover rate, 51
advocacy journalism, 156
AFL-CIO, 5, 141, 155
Amalgamated Meat Cutters and Butcher Workmen of North America, 16
ammonia gas, 29
Anderson, A. D., 16
Andrews, Robert E., 205–206
anti-union activities: business leaders urging no vote, 64–65, *65*; "captive audience meetings," 55–56, 223, 228; firings of black union supporters, 75–76; identifying and monitoring union supporters, 53; law firm and personnel directing, 41–42; Luter on, 100–101; pitting blacks against Latinos, 75; union-busting businesses, 2, 42, 74, 100–101
Arce, Carlos Noel (Father Carlos), 186–191, *194*, *195*, 201
Armour brand, 10, 16–17
Arnold, Remmie L., 12
attorney/client privilege issue, 119
automation, 16–17, 23

back pay awards, 3, 131, 162–164, 204, 232, 234
bacterial groundwater contamination, 31
Bailey, C. J., *199*
Bailey, Lenora, *199*
Banks, Delilah, 70, *71*
Barber II, William, 174, *175*, 212–213
Barr, Matthew, 274

Barrett, William P. "Bill": background, 43, 44–45; banned from anti-union meetings, 85; Buffkin's testimony against, 118–120, 124–125; laughing at anti-union vote, 95; and McMillan's firing, 90, 105–106; and NLRB case, 113–114, 116; as onsite labor attorney, 41–43, 113; replaced by Robertson, 137; resisting handing over documents, 116; statements to workers, 84
Barry, Dave, 100–102
Battista, Robert J., 137
Bell, Matthew, 196
black community beef boycott, 167
black workers: after RICO suit, 226; casings area, 38, 64; and Father's Day protest, 171; and Martin Luther King Jr. Day, 192, *193*, 227–228. *See also* race issues at Smithfield Packing
Bladen County, 9, 26, 29, 65, 67, 69–70, 72, 145
Blair, Kevin, 145
Blakey, G. Robert, 219
blast-chill wind tunnel, 38, 69
*Blood, Sweat, and Fear* (Compa), 143
Blount, James, 96
Border Protection, Anti-Terrorism and Illegal Immigration Control Act of 2005 (H. R. 4437), 159
"Boss Hog" series (*The News & Observer*), 26, 32, 274
Boston, Massachusetts, 173
Bowser, Renee, 113–114, 125–127, 131, 138
box warehouse, 39

boycott, 167, 172–175, 185, 217, 228

brains: market for, 39

Brazilian market, 69

Brown, Jasper: as African American in courtroom, 113; background of, 111; on cost of trial, 113; cross-examining Luter, 127–128; filing EEOC complaint, 111; on Judge West, 114; as lead attorney for NLRB, 108, 111–112; lunching with other attorneys, 116; on NLRB's slowness to respond, 112; putting Buffkin on stand, 107–109, 117–118, 126, 132; on Smithfield case, 111, 163

Brown, Joe, 169

Brown, John, 154

Bruskin, Gene, 150, 158; background, 149–150, 151, 155; on Border Protection Act, 159; card check versus secret-ballot method, 158, 164, 206, 210, 215; celebrating 2008 union vote, 229; comparison with Joe Luter, 151; departure after RICO dismissal, 226–227, 238–240; and Employee Free Choice Act, 152, 210; on Father Carlos, 189, 194; on free speech, 220; and Harris Teeter Rancher campaign, 167–173, 176–181; included in RICO suit, 217–224; on Joe Luter, 224–225, 238; and Keith Ludlum, 183; on Latino May Day demonstration, 159–161; and Latino walkout, 186–187, 191, 194; Leila McDowell and Justice@ Smithfield, 154–155, 157; linking black and Latino rights, 165; and Martin Luther King Jr. Day, 192; on negotiations as stalling technique, 215; on NLRB, 151–152; on secret executives UFCW negotiations, 211–212; targeting shareholder meetings, 210–214; as union campaign director, 150, 152–153; on "yellow fever" day, 165. See also Change to Win

Buffkin, Davie, 20, 77, 104, 118, 234

Buffkin, Nicole, 20

Buffkin, Sherri Wright, 36; private life: 35, 104; on Ada Perry's death, 234; anonymous letter given to husband, 117–118; birth and childhood, 19–20; on daughter's feelings about her work, 109; on Joe Luter, 235; life after Smithfield, 231; at work: at 1997 union vote, 93–94, 96; and Ada Perry, 89, 102–104; anonymous letter about, 117–118; attempting to fire employee, 52; attempting to stay neutral, 87–88; at Aug. 22 meeting, 93–94, 96; contacting Jasper Brown, 107–109; as crew leader, manager, 39, 52, 77; filing postdated warning to McMillan, 106; fired, 106–107; and Larry Johnson, 39, 78, 87, 105, 106–107; and McMillan, 85, 90, 106; Motrin issue, 122, 124–125; and Null, 78–79, 89, 104–107; opposing unionization, 60, 77, 79, 108; ordered to falsify records, 106; ordered to fire employees, 120; performance reviews and top raises, 78; and Perry's firing, 102–104; on Priest as "thug," 41; as production support manager, 77; and profanity in plant, 63, 79; as purchasing agent, 104; receiving gifts from salesmen, 104–105; reporting Null's harassment to Johnson, 105–106; setting up for meeting, 88–89; and Susie and Billy Jackson, 77–78, 104, 107, 121, 234; suspended while on vacation, 106; at Tar Heel plant opening, 35; testimony before U.S. Senate, 136–137; umbrellas and phone issue, 106, 122–124, 126; on union, 75, 108; on use of "nigger," 80; at trial: agreeing to testify, 108; anonymous letter before testimony, 117–118; character witnesses against, 121–123; cross-examination, 120–121; Judge West on,

132; perjury issue, 120; preparation for trial, 116–118; testimony at trial, 114, 118–120, 124–126
Burlington Industries, 147
Burr, Richard, 161
bursting hogs, 76, 114
Bush, George H. W., 220
Bush, George W., 44, 203, 223

Cahoon, Lawrence B., 29
campaign contributions, 26, 32–33, 181
Campbell, Mac, 83
"captive audience meetings," 55
card check versus secret ballots, 158, 164, 206, 210, 215
Carolina Cold Storage, 69
"Carolina Crowd," 58
Carolina Mills, 147
Carroll's Foods, 26
casings area, 38
Change to Win, 141–142, 148, 150, 217. See also Bruskin, Gene
Charlotte Observer, 228
Chavez, Cesar, 160
Chavez-Thompson, Linda, 86
Chelsea, Massachusetts, 173
Chinese market, 1, 39, 69
Circle Four Farms, 138
civil rights trial, 133–134
Clean Water Act (federal), 71
Clinton, Bill, 203
Clinton, Hillary, 222
Coble, Howard, 39
collective bargaining: Liebman on, 203–205
Columbus County, 145
communism fears, 12, 22–23
Compa, Lance, 143–144, 162
company owned farms, 10
compensation claims, 64, 121–122, 143–144, 155–156, 167, 222
Congress Cooks: pork brains recipe, 39

consumers: as campaign focus, 142
contract farming, 10, 28–29, 64, 237
conversion department, 36, 38, 69, 75
conversion laundry room, 79, 87
conveyor belts, 35–37, 39
corn, 27, 72
corporate America, 4, 207
"corporate campaign": on human rights, 149–158
cotton mills, 19, 21, 23
Council, Chris, 66, 75–76, 114, 131, 163
Cromartie family, 146
Crump, Joe, 153–154
Cudahy Packing Company (Patrick Cudahy), 16, 18, 39–40, 60–61, 100
Cumberland County, 145
Curtatone, Joseph, 172
cut floors, 35–36, 38–40, 64

Davis, Tara, 131, 163
"Day of Action" events, 167, 169, 170
"Deborah from Greensboro," 179
Deen, Paula, 176–181, 185, 217, 219
dehairing, 36
Diane Rehm Show, 178–179
Dickerson, C. Wyatt, Jr., 13
Dole, Elizabeth, 161
dry kill floor, 36–38
Duplin County, North Carolina, 26, 29–31

economic polarization, 4
Edwards, John, 136
"Eight-A-Ones" violations, 57
Ekrich brand, 10
Employee Free Choice Act, 152, 204–206, 208, 210
English, Dan, 146
environmental issues in North Carolina, 11, 26, 29–33
Environmental Protection Agency (EPA), 72, 128
equal pay for unequal work, 64

Families Against Mandatory Minimums, 155
family farms, 27–28
Family Fish House, 14
family leave law, 145
farm subsidies, 27
farmers: individual contracts with, 10, 64
Farmland Foods brand, 10, 139
Father Carlos, 186–191, 194, 195, 201
Father's Day protest, 170–171, 218
Fayetteville, North Carolina, 9, 176
federal air traffic controllers strike, 62
federal family leave law, 145
"Feeding the Hungry" program, 224
feijoada, 69
Fiedler, Jeff, 153
Field, Sally, 108
First Amendment versus RICO, 218–219
Floyd, Erica, 161
Food, Inc. (video), 46–47, 198–200
Ford, Gerald, 14
foreign competition, 23
foreign market: for pork, 85
Forrest, Anthony, 91–92, 96–97
Foxx, Virginia, 207
freedom of association, 143
Freeman, Richard, 5

"gag order" in RICO settlement, 232, 256n1
gams, 37
gassing of hogs, 47–48
Gattalaro, Don, 113–114, 119, 120–122, 132
Getman, Julius, 221
Gheen, William, 195
Gilliard, Sherman, III, 55, 82–83, 115
Glover, Danny, 214, 233
Goldsboro Hog Farms, 26
Goldsmith, Willis H., 134
Gone Hog Wild (newsletter), 227
Grice, Robert, 161
Guadalupe: story of, 196

Gwaltney, P. D., Jr., 11
Gwaltney plant, 10–12, 14, 71

Hacker, Jacob S., 6
Hall, Ray, 145
Hampton Institute, 111
handbilling: at plant entrance, 53–54, 75–76, 84, 91–92, 160, 177
"hangman philosophy," 60
Hansen, Joe, 223–224
Harbrant, Bob, 153
Harkin, Tom, 136
"Harris-Teeter Rancher" campaign, 167–173, 185, 217–218
Hearst, Anne Randolph, 238
helmets/hard hats: color coding of, 38; stickers/writing on, 184, 227
Herbert, Bob, 156
Herrera, John, 161
hog farms, 10, 19, 27–32, 41, 70
hog openers, 75–76
hogs: mistreatment of live, 45–48; processing of at Smithfield, 35–39; workers treated same as, 46
hogs' heads, 35, 37–39, 47, 64, 84
"hog-whisperer," 45
Holland market, 69
hooking hogs, 35–37, 145, 225
Hormel plant P-9 strike, 61–63
hotdogs, 38
House Bill 467 (NC Senate), 31–32
human resources: erratic record-keeping in, 51
Human Rights Watch (HRW), 143, 156, 162, 167, 173
Hunt, Oliver "Ollie," 183

IBP: competition from, 25
ICE: creation of, 158–159; IMAGE program, 185, 191, 194–197; Smith & Company raids, 192–193
immigrants. See Latino workers; undoc- umented workers

independent contractors, 204

injured workers: court testimony from, 116; due to high line speeds, 143; due to inadequate training, 63; firing of, 64, 143–144; forced to continue working, 48; increased injuries at Tar Heel, 61, 174; Joe Luter IV on, 162; letter to Paula Deen on, 180; in meatpacking industry, 10, 155; mistreatment of, 48, 143; OSHA reports on, 177; RAA publication on, 167; Ray Hall suit, 145; repetitive-motion injuries, 143, 180; settling compensation cases, 144; Smithfield desire to avoid court over, 222; union protection for, 15–16, 143, 174

inspecting the line, 37

Isle of Wight County, 25

Jackson, Billy, 77, 79, 85, 90

Jackson, Jesse: Buffkin avoiding rally, 87; and Chris Council's firing, 75; having worked in packinghouse, 91; National Rainbow Coalition, 149; Perry attending rally, 87; at plant on voting day, 91; at pro-union events, 65–66, 66, 82, 85, 97; and racist graffiti, 82; signing petition, 214; speech and prayer by, 86

Jackson, Susie, 77–78, 106–107, 121, 123, 126, 234, 255n4

jamming the line, 46

Japanese market, 69

John Morrell & Co./John Morrell Group, 10, 100, 138, 236

Johnnie's Foodmaster, 172

Johnson, Larry: Brown's testimony against, 118; and Buffkin, 39, 78, 105–106, 119–120, 235; clarification letter about union stickers, 184; court testimony of, 116, 121–122; during walkout, 186; and EEOC lawsuit, 135; gifts from salesmen, 105; home of, 146, 235; and Jesse Jackson, 91; Judge

West on, 131; and Lawanna Johnson, 55; leaving union for management, 39–40, 84; and McMillan, 90, 106, 119; meeting with Father Carlos, 190–191; and Perry, 87, 103, 119; post-settlement remarks by, 235–236; promoted after trial, 138, 258n14; and Spann, 55; at union vote, 92, 94, 96, 99; Ward remarks on, 84

Johnson, Lawanna, 54–55, 59, 81, 114, 131, 163

Johnson, Nelson, 63, 167–171, 168, 196–197, 212, 214

Jones, Larry Charles, 76, 114, 131, 163

Jungle, The (Sinclair), 1, 157, 219

Justice for Janitors campaign, 141

Justice@Smithfield campaign, 193, 198; Bruskin and, 154, 239; "Day of Action" events, 167, 169, 170; disorganization within, 159; going national, 239; "Harris-Teeter Rancher" campaign, 167–173, 185, 217–218; Leila McDowell and, 154–157, 170, 173, 178, 180; and May Day march, 159–161; not tied to union issue, 169–170, 173; providing example letters on website, 175; RICO suit against, 217–218, 224; targeting shareholders, 210–211; t-shirts, 159, 165, 168, 198, 212, 214, 239; website, 175, 218, 224, 239

Katz, Joel, 119–120, 126

Kennedy, Edward, 136–137, 208

Kennedy, Robert F., Jr., 28, 29–31, 33

Kerry, John, 141

kidneys: "popping" of, 38

kill floor/killing room: Ada Perry incident with workers from, 103; animal abuse on, 47; blast-chill wind tunnel added to, 69; closed during raid, 194; council distributing union fliers on, 75; dry kill workflow, 37–38; Fat-O-Meter, 81; generally black men working on,

64, 75; hog behavior on, 46; injured
worker on, 48; Joe Luter working on,
13; Johnson addressing, 212; tempera-
ture of, 36; union support on, 227;
Ward working on, 83–84; workflow
on, 35
King, Larry, 178, 179
King, Martin Luther, Jr., 160, 192, *193*,
*198*, 227–228
King Kullen, 175
Knight, Thomas, 22
Korean market, 39, 69
Krugman, Paul, 4
Ku Klux Klan, 12, 22–23

labor law: "ossification" of, 4; union busting as "labor law," 42; weakness of, 6
lagoon farms/manure pits, 29–32, 70
Latino workers, 147–148, 225; after
RICO, 225; August 2007 raid, 197–
201; and blacks, 74–76, 133, 136, 142,
227–228; cleaning crew walkout,
plant closure, 194–195; Compa report,
speech to, 143–144; in competition
with blacks, 3, 76, 142; Dan English
on, 146; "Day without Immigrants,"
160–161; depressing wages, 73–74;
and Father's Day protest, 171, 218;
*Food, Inc.*, 198, 200; and ICE IMAGE
program, 185–186, 191, 193–195, *199*,
201; illiteracy among, 74; language bar-
rier, 146; May Day demonstrations by,
159–161; November 2006 walkout by,
185–192, *189*, 227; Peña recruitment of,
142–145; reviving and changing small
towns, 147–148; Swift & Company
raid, 192–193; ten percent in 1994, 51;
union assistance to, 141–142; union recruitment of, 74, 85–86, 157, 159, 185,
189; working in cut and conversion,
64, 75. *See also* Justice@Smithfield
campaign; Ludlum, Keith Alan; race

issues at Smithfield Packing; RICO
lawsuit; undocumented workers
Lauritsen, Mark, 195
LeDuff, Charlie, 147, 274
Lendon, Patsy, 81–82, 114–115, 131
Lewis, Edward Ross, 49
Liberty Equities, 13–14
Lichtenstein, Nelson, 6
Liebman, Wilma, 3–6, 44, 137, 203–205
line speeds, 10, 16–17, 37, 47, 57, 63–64,
143, 155
livestock section, 36, 41, 58, 174, 183–184,
208–209
lobbying: free speech as racketeering, 219
longleaf pine forest, 21
Lowrey, Joseph, 86
Ludlum, Dilcia, 240
Ludlum, Keith Alan: on 2008 unionization vote, 229; agreement to leave Tar
Heel plant, 226; background of, 41;
currently unemployed, 233, 240; on
Employee Free Choice Act, 205, 206;
firing of, 59, 114, 210; on hangman
philosophy, 60; during ICE raids,
198–200; job in Livestock, 45–46,
164; Livestock water issue, 208–209;
objecting to labor law violations,
49; observing management cruelty,
48; provoked into fight, 209–210;
reinstated after verdict, 163–165; on
secret-ballot process, 207; and television campaign, 174; testifying before
Congress, 206–207; as union local
president, 232–233; union more cau-
tious than, 184–185; as union recruiter,
48, 53, 58–59, 183; "Union Time" on
hard hat, 184; on winning battles, 241
Lumbee Indians, 40, 51
Lumberton, 160–161, 200, 229, *229*
Luter, Barbera Thornhill, 14, 15, 18
Luter, Erika, 237
Luter, Joseph Williamson, 11

Luter, Joseph Williamson, Jr., 19

Luter, Joseph "Joe" Williamson, III: alliance with "The Circle," 26; as "boss hog," 15; Gene Bruskin on, 224–225, 238; Larry Johnson on, 235; life in retirement, 237–238; losing, buying back company, 13–14; personal life of, 11, 13–15; reasons for choosing North Carolina, 19; resignation as CEO, 162–163, 211; salary of, 127, 146; Sherri Buffkin on, 235; strike after pay reduction by, 18; and Tar Heel plant's hard line, 144; testifying at trial, 126–129; vertical integration at Smithfield, 25; working in slaughterhouse, 13; statements by: denying ordering records destroyed, 72; on environmental laws, 128; on NLRB, 127–128; on small farms, 27; on unions and union votes, 10, 18, 100–102, 129, 214, 228–229; on vertical integration, 27; on wastewater treatment case, 72

Luter, Joseph W., 13, 162, 235

Luter, Karin, 237–238

Luter, Laura, 13

Luter, Leigh, 13

Luter Packing Company, 11

"Luter the Polluter," 32

lynching threats, 12

Maupin Taylor Ellis & Adams, 41–42

May Day 2006 march, 160–161

McAfee, Gail D., 173

McCarthy, Joseph, 22

McDonald, Fred, 65, 75–76, 114, 131, 163

McDowell, Leila, 154–157, 170, 173, 178, 180

McGinnis, Mike, 125

McInerney, Jay, 238

McIntyre, Mike, 161, 195

McMillan, Margo: and Buffkin, 88; Buffkin filing postdated warning to,

106, 108; Buffkin testimony regarding, 119–120; court awarding reinstatement and back pay to, 131, 163; firing at issue in trial, 114–115; Jackson testimony regarding, 121; Johnson testimony regarding, 123; as laundry crew leader, 79, 85, 90, 105–106; Null testimony regarding, 124; objecting to uneven discipline, 80; petition from coworkers, 85, 90, 123; transferred then fired, 90; union activity of, 79, 86–87; warned about union activity, transferred, 85

McNair house, 146

meatpacking industry, 10, 15–18

Mendez, Pedro, 185

Menino, Thomas, 173

Mexico and NAFTA, 72. See also undocumented workers

middle class: hollowing out of, 4, 207

Millan, Antonio, 159

Miller, George, 152

Mitchell, Bennie, 157

moral victory: as extortion, 220

Morganthall, Frederick J., 170–172

Morrell Group, 10, 100, 236

Morris, Henry L., 51–52, 56, 60, 67, 70, 88

Mother Jones, 181

Motrin issue, 122, 124–125

Moyer Packing, 139

Murphy, Wendell H., 26, 28, 32

Murphy-Brown LLC, 31

Murphy Farms, 69, 138

Mutual Agreement between Government and Employers program. See ICE

NAACP, 143–144, 168–169, 174–175, 175, 212–213

NAFTA: effects of, 60, 72, 158

Nash, Vincent, 177

Nathan's brand, 10

National Baptist Convention, 157

national income distribution, 3–4

National Labor Relations Act, 2–3, 43–45

National Labor Relations Board (NLRB), 3; administrative law judges, 44; called a failure, 7; drop in case load, 44; Employee Free Choice Act alternative to, 152; hearings process, 113; injunction request denied (1998), 105; investigative arm of, 44; Liebman articles on, 3, 203–205; presidential political appointees, 43; sanctions against textile industry, 22; Senate filibuster of appointments to, 44; staggered terms on, 44; "toxic" in South, 5; UFCW complaints, 57–59, 105, 135; union loss of faith in, 4; weak and slow enforcement by, 2, 44, 151

Native Americans, 20, 40, 51

New Deal, 203

*New York Times, The*, 156

*News & Observer, The* (Raleigh), 26, 29, 32, 169, 274

"nigger": use of term, 63, 80

Niman, Nicolette Hahn, 30, 275

nitrogen, 31

Nixon, Richard, 14, 17

no-match letters, 185–186, 188

*Norma Rae* (film), 108

North Carolina, 25; as a "company town," 33; legacy of slavery in, 21, 35, 60, 146; Luter's reasons for choosing, 19; poverty in, 20; state government, 26

Null, Jere T.: after leaving Smithfield, 235–236; Buffkin fired on order of, 106–107; Buffkin personal relationship with, 78–79, 104–106, 108; Buffkin professional relationship, 89–90, 103, 105; Buffkin sexual harassment suit against, 132–133, 135, 255n4; Buffkin's testimony regarding, 118–119, 126, 134; as general manager at Tar Heel, 78; on Japanese market, 70; on Jesse Jackson, 91; Judge West on, 131–132; ordered by judge to read election notice, 135; and Perry, 87–89; post-settlement, 236; promoted after trial, 138; promotions, 78; regarding Ward and Rodriguez, 134; remarks to Chad Young, 92; and Susie Jackson, 121; as technical service director, 40; testimony at trial, 116–121, 123–124; videotaped speeches, 134; Ward on, 84

Obama, Barack: Bruskin photo with, 239; campaign endorsements for, 181; on Employee Free Choice Act, 208; NLRB appointees under, 5, 44; pro-union statements by, 222–223, 228

one-percent richest, 3–4, 205

O'Neill, Pat, 218, 223

Onslow County spill, 70

"Operation Dixie" union drive, 21

Operation Wagon Train, 192–193

*Oprah Winfrey Show, The*, 180–181

P-9 strike at Hormel, 61–63

"packaged with abuse" campaign (RAA), 155, 167, 217

Packland Holdings, 139

Pagan River, 32, 71

Palmer, Alberta, *199*

Papian, Jim, 139

Patrick Cudahy Packing Company, 16, 18, 39–40, 61, 100

Payne, Robert E., 220–221

Pelham, Otis, 82–83

Peña, Eduardo: at 2008 vote count, 228; as campaign manager, 142–145, 158, 185; and family leave law, 145; and *Food, Inc.*, 198, 200; and ICE IMAGE program, 185–186, 194–195, 198; and Latino walkout, 187–188, 191; RICO suit against, 218; on union and trust, 196

People for the Ethical Treatment of Animals (PETA), 45–46

Perry, Ada "Grandma": Buffkin on death of, 234; court awarding reinstatement and back pay to, 131, 163, 234; fired by Hall, 103–104, 114; job and job performance, 79–80, 103, 105–106; trial testimony regarding, 114–115, 121, 123–124; trial testimony/statements by Buffkin on, 89, 103, 105, 108, 109, 119–120; as union observer, 86–88, 92–95, 97, 102–103
phosphorus, 29, 31, 71
picket lines: crossing of, 61
Pierce, Gorrell, 22
Pittman, Dennis, 169–170, 187, 190, 194
plant closure: due to ICE raid, 194–195
police: company, 41, 131, 133–134, 138, 142, 145, 162; county, 12, 98, 148, 161, 171–172, 213. See also ICE
Pope, Larry, 173, 214, 225
"popping kidneys," 38
postdated warning letters, 106, 108
"Pressure Is On, The: Organizing without the NLRB" (Crump), 153–154
Prestage Farms, 26, 236–237
Pridgen, Timm, 54
Priest, William Daniel "Danny": assaulting Ward, 97–99, 117; background of, 40; civil suit against, 132–133; harassing of union workers by, 45, 53–54; hired as director of security, 40–41; and Jesse Jackson, 91; made chief of special police, 138; at union vote count, 94, 97–99
prison inmates as workers, 52, 54
prodders, 35, 46
"property": union as tangible, 220–221
pullers, 38

race issues at Smithfield Packing: accusations of racial favoritism, 145–146; blamed on unions and communism, 22–23; company supported by white government, business, 63; division of labor by race, 51, 64; earliest workers mostly black, 35; fewer Latinos after RICO suit, 225; hog farms in black population areas, 31; Latino workers lowering wages, 73–74; management encouraging racial animosity, 60, 74–75; mistrust because of, 146–147; post-RICO black/Latino solidarity, 227–228; union support by race, 91; Ward on, 99; white supervisors, black workers, 39, 51, 63; "yellow fever" day with blacks and Latinos, 165. See also Latino workers
racketeering lawsuit, 217–227, 232, 256n1
Reagan, Ronald, 6, 17, 62, 204
Reconstruction, 23
Red Scare, 12, 22–23
Rehm, Diane, 178–179
religious organizations' activism, 157, 173–174, 177
rendering, 39
Republican Party, 5
Research Associates of America (RAA), 153, 155–156, 167, 173, 217, 222
Richter, George H., 223
RICO lawsuit, 217–227, 232, 256n1
Righteous Porkchop (Niman), 30, 275
right-to-work movement, 3, 6, 19, 22, 62, 232
"Roberto's garage," 189
Robertson, Greg, 137–138
Robeson County, 9, 29, 54, 73, 145–147, 191, 197
Rodriguez, John Renee "Johnny," 98, 99, 117, 132–134
Roosevelt, Franklin D., 2, 43
Ross, Tom, 101–102

safety campaign, 154. See also Justice@ Smithfield campaign
Sampson County, 29–31
Sanders, Bernie, 6
Sarandon, Susan, 214

Saunders, Lee, 5
sausage, 38
Schaumber, Peter, 44
Schellpeper, Timothy O., 223
Scott, Tony, 94–95
sealed depositions, 135
security guards: at Smithfield, 41
self-organizing, 188
Sensenbrenner, Jim, 159
Service Employees International Union
    (SEIU), 141, 181
shacklers, 47
sharecroppers, 19, 21, 23, 28, 86
shareholders: as campaign focus, 142,
    210–214
Sharpton, Al, 214
shipping and receiving section, 36, 39
Shuanghui International Holdings Ltd., 1
Shugrue, Vicar Monsignor Michael P.,
    187
Simpson, George, 55–59, 114, 131, 163
Sinclair, Upton, 1, 157, 219
skull, 39
Slaughter, Terry, 183, 209, 214
slaughterhouse work, 9–10, 35–39
Smith, Dale "Big Country," 97
Smithfield Beef Group, 139
Smithfield Foods: acquisitions by,
    138–139; brand tied to locale, 9;
    company owned farms worldwide, 10;
    fined for polluting river, 71; investment
    in North Carolina pig farms, 10, 25;
    lowering costs as primary goal, 2;
    restructuring of, 13–14; as world's
    largest hog producer, 138
Smithfield Ham versus Smithfield ham, 9
smithfieldjustice.com website, 170, 172,
    224
smithfieldjustice.org website, 239
Smithfield Packing (Tar Heel, NC),
    65; animal mistreatment at, 46–48;
    Change to Win campaign, 141–142,
    148, 150, 217; compared to other

Smithfield plants, 139, 143–144; export
    processing room, 70; initial legal team,
    41–42; initial management staff, 39–41;
    injuries and line speed issues, 63–64;
    New York Times on, 156; NLRB
    charges against, 105, 112, 163; not fined
    for labor violations, 207; plant open-
    ing, 10, 35; size of, 9, 70; a subsidiary
    of Smithfield Foods, 10; suing state
    over wastewater permit, 70–71; taxes
    to Bladen County, 69; on unfair labor
    practices claims, 135; union recruit-
    ing Latino workers, 73–74; wages at,
    83, 113, 139; workflow at, 9–10, 36,
    45–48; "yellow fever" day at, 165. See
    also Justice@Smithfield; race issues
    at Smithfield Packing; trial against
    Smithfield Packing
snout trimmers, 64
social compact/contract: violation of, 62,
    204
Social Security numbers, 200–201
Somerville, Massachusetts, 173
Southeast Day of Action, 170
Southern Environmental Law Center, 71
Southern Faith, Labor, and Community
    Alliance, 167–169, 168
Southern States Industrial Council, 12
soybeans, 27
Spann, Gregory, 54–55
state minimum-wage laws, 6
Stewart, Emily, 226
"sticker" job, 47
St. Pauls Cotton Mill Co., 147–148
strikes, 6, 61, 204
stunner tool, 47
Sutton, Crystal Lee, 108
Sweeney, John, 141
Swift & Company raids, 192–194, 196

Taft-Hartley Act, 22
Tar Heel, North Carolina, 20, 30; anti-
    union signs in, 83; hog farms, 10, 19,

27–32, 41, 70; union trailer in, 53–54, 67, 82, 86, 142

Tar Heel slaughterhouse/Tar Heel plant. *See* Smithfield Packing (Tar Heel, NC)

Teamsters' UPS strike, 141

Teeter, Harris, 167–173, 185, 217–218

Terry, Paul, 147–148

textile industry, 22, 23

Thornhill, Barbera, 14, 15, 18

Tienda La Hacienda, 228

TIPS (threaten, interrogate, promise, spy), 42–43

trial against Smithfield Packing, 105; reasons for delay in filing, 112; list of charges, 112, 114; judge and attorneys in, 113; start of, 114; Smithfield's failure to produce records, 115–116; daily conduct, mood of, 116–117; Young and Ward testimony, 117; Buffkin testimony, 117–121; Null's testimony, 117–121, 123–124; Buffkin recall, 125–126; Luter's testimony, 126–129; ruling, 131–132; verdict stayed, overturned, 132–134; appeal to full board, 134; original judgment upheld, 137; appeal to federal circuit court, 137, 162; verdict upheld in federal circuit court, 161–162. *See also* Brown, Jasper; civil rights trial; West, John H.

Troy, Shawn, 81

trucks for hog transport, 45

Trump, Donald, 5–6

t-shirts: anti-union, 65; pro-union, 159, 165, 168, *198*, 212, 214, 239

turnover rate and absenteeism, 51–52

turpentine, 21

twenty-fifth house speeches, 134, 223

two-chain kill plants, 45

Tyson Foods, 27

UFCW (United Food and Commercial Workers International Union), 2, 53; calling off Hormel strike, 61–62;

Change to Win Campaign in Tar Heel, 142; endorsement of Obama, 181; and Hormel P-9 strike, 61; NLRB complaint filed by (1993), 57–58; NLRB complaint filed by (1998), 105; and Oprah Winfrey, 180–181; RICO lawsuit against, 217–227, 232, 256n1; secret negotiations, 211–212; sufficient cards to hold union vote, 59

umbrellas: and phone issue, 106, 122–124, 126

undocumented workers: arriving after NAFTA, 72–73; demonstrations in support of, *198*; ICE raids and arrests, 185, 191, 194–201; and immigrant rights movement, 160–161; influx of after NAFTA, 60, 72–74, 76, 146–147; intimidation of, 3; no-match firings of, 185–186, 188; threatened with deportation, 131. *See also* Latino workers

union-avoidance consulting, 42

union headquarters: union trailer in Tar Heel, 53–54, 67, 82, 86, 142; workers center in Red Springs, 142–143

"Union Man" Ward as the, 84

union observers, 87, 92–95, 97, 102

union organizers, 141–148, 189

union petition (2007), 214

unions in United States: Amalgamated Meat Cutters and Butcher Workmen of North America, 16; card-check system, 152, 158, 164, 207–208, 214–215; National Labor Relations Act, 2–3, 43–45; and political parties, 141; recent drop in membership, 4, 6; recent drop in strikes, 6; in the south, 21–23; split over strategies, 141–148; union-busting (*See* anti-union activities)

union vote (Aug. 1994), 59, 64–67, 86, 100–101, 137

union vote (Aug. 1997), 13, 82, 86, 91–96 (*96*), 137; violence following, 95–100 (*99*)

union vote (Dec. 2008), 228
union vote off-site venue issue, 164
United Packinghouse Workers of America, 11–12
UPS strike, 141
U.S. Courts of Appeals, 161–162, 211, 220
USDA inspections, 37–38, 76, 194, 207
U.S. Pure Food and Drug Act, 1
U.S. Supreme Court, 6

Vazquez, Margarita, 188
vertical integration at Smithfield, 25

wages: deducted from monetary penalties, 2; minimum wage, 6, 9, 20, 89; national stagnation in, 6; for slaughterhouse workers, 15
Walker, Leonard, 192
Walsh, Dennis P., 137
Ward, Rayshawn: at anti-union meetings, 84–85; assaulted, arrested during vote count, 97–99; background of, 83; on Barrett, 42; compensation voided due to waiver, 133; fired, 99–100, 114–115; judge's verdict and statement on, 131–132; physical description of, 41, 42, 98; signing waiver of liability, 99; testimony at trial, 117, 123; as union activist, 84; union observer at vote count, 93, 94–95; verdict upheld, 163; winning civil suit, 132–134; work at Smithfield, 83–84
Ward v. Smithfield, 252n33
wastewater permit issue, 70–71
Waterkeeper Alliance, 31

water pollution: from hog farms, 29–32, 70–71
Wellstone, Paul, 136
West, John H.: background of, 113; courtroom demeanor of, 113–115; termination records issue, 115–116; attorney/client privilege issue, 119; during Buffkin's testimony, 120, 122; direct questioning of Johnson, 122–123; on recalling Buffkin, 125; during Luter's testimony, 126–129; ruling, 131–132, 138, 164; Smithfield's appeal of ruling, 134, 162; verdict mostly upheld by NLRB, 137, 164, 220; bias charges against, 137; Buffkin on, 234; Johnson on, 236
wet kill floor, 36–38
wheat, 27
white workers. *See* race issues at Smithfield Packing
Williams, Pamela, 81
Wilmington, North Carolina, 20
Winfrey, Oprah, 180–181
*Winston-Salem Journal*, 197
worker injury issue, 143
workers' compensation, 64, 121–122, 143–144, 155–156, 167, 222
*World, The*, 196

"yellow fever" day, 165
Young, Chad, 91–92, 94–98, 117
"Young Blood," 84

zoning, restrictions against, 26